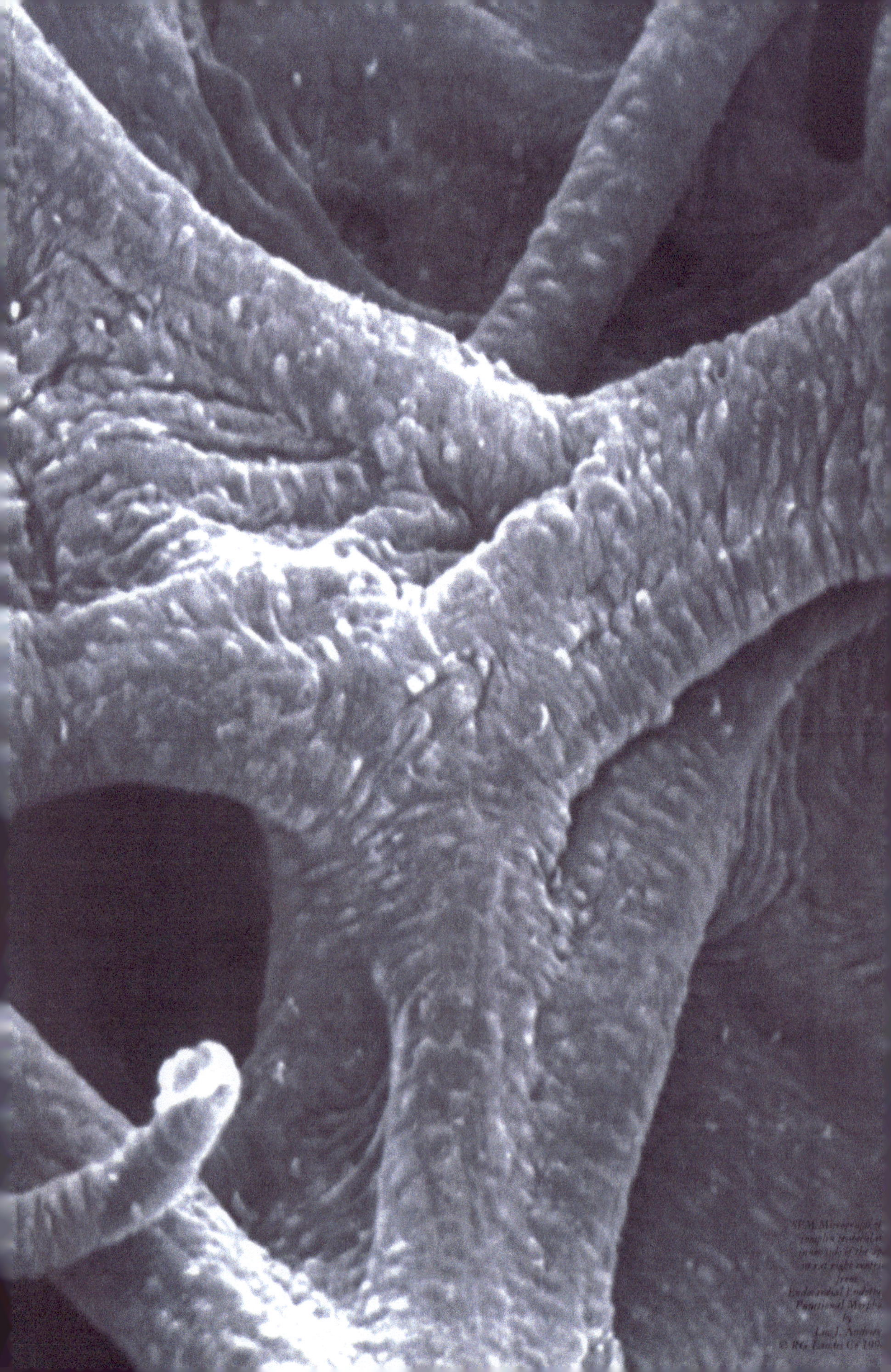

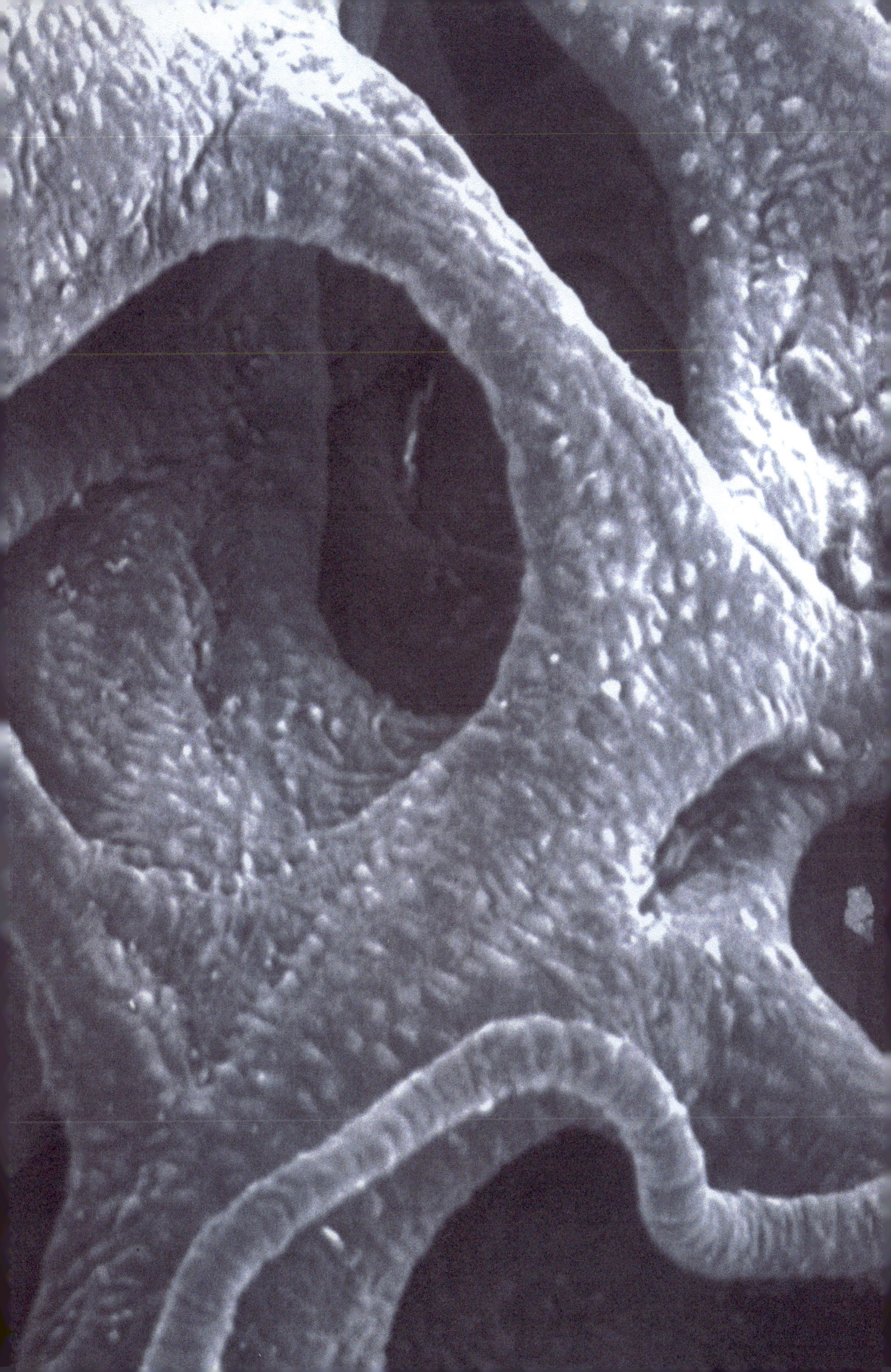

MEDICAL
INTELLIGENCE
UNIT

c-MYC FUNCTION IN NEOPLASIA

Chi V. Dang, M.D., Ph.D.
Linda A. Lee, M.D.
The Johns Hopkins University
School of Medicine
Baltimore, Maryland, U.S.A.

Springer-Verlag Berlin Heidelberg GmbH

Medical Intelligence Unit

C-MYC FUNCTION IN NEOPLASIA

ISBN 978-3-662-22683-4 ISBN 978-3-662-22681-0 (eBook)
DOI 10.1007/978-3-662-22681-0

Originally published by Springer-Verlag Berlin New York Berlin
Heidelberg Lndon Paris Tokyo Hongkong Barcelona Budapest in 1995
Softcover reprint of the hardcover 1st edition 1995

Publisher's Note

R.G. Landes Company publishes five book series: *Medical Intelligence Unit, Molecular Biology Intelligence Unit, Neuroscience Intelligence Unit, Tissue Engineering Intelligence Unit* and *Biotechnology Intelligence Unit.* The authors of our books are acknowledged leaders in their fields and the topics are unique. Almost without exception, no other similar books exist on these topics.

Our goal is to publish books in important and rapidly changing areas of medicine for sophisticated researchers and clinicians. To achieve this goal, we have accelerated our publishing program to conform to the fast pace in which information grows in biomedical science. Most of our books are published within 90 to 120 days of receipt of the manuscript. We would like to thank our readers for their continuing interest and welcome any comments or suggestions they may have for future books.

Deborah Muir Molsberry
Publications Director
R.G. Landes Company

Dedication

C. V. D. wishes to dedicate this monograph to his parents, wife — Mary, and his children Eric and Vanessa, who each inspire compassion and curiosity.

CONTENTS

PREFACE

The c-*myc* protooncogene is frequently altered in many human cancers and the understanding of its function cannot be overemphasized. This monograph is inspired by an explosion of recent discoveries that have shed light on the function of c-Myc in neoplasia. This inspiration emanates not only from the beauty of these discoveries but also specifically from those who made the discoveries. The first two chapters are brief overviews of our current understanding of signal transduction and neoplastic transformation. These chapters place in context the role of c-Myc in the complex molecular interactions that regulate cell proliferation. Subsequent chapters review the discovery of *myc* starting with the recognition of retroviral *myc* and the cloning of the vertebrate homologs. Regulation of c-*myc* gene expression and the properties of its protein product are discussed. An overview of c-*myc* in human neoplasia is then presented. The remaining chapters highlight current observations on c-Myc protein structure, function and its interactions with other proteins. The final chapter gives an overview, while still incomplete, of genes that appear to be regulated by c-Myc and their role in c-Myc induced neoplastic transformation and apoptosis. In gathering the information for this monograph, only selected references could be included because of the allocated space and time, and we therefore apologize to our colleagues for the omissions.

Acknowledgment

Orginal research from our laboratory discussed in this monograph was in part supported by the National Cancer Institute — National Institutes of Health, American Cancer Society, American Gastroenterology Association and the Wilbur-Rogers Foundation. C. V. D. is a Scholar of the Leukemia Society of America. L. A. L. is a A. Guerrieri Clinician-Scientist.

CHAPTER 1

INTRODUCTION

1.1 SCOPE OF BOOK

An explosion of novel findings in the past decade has contributed to the great progress toward understanding the biology of human cancers. Much of this progress can be attributed to our ability to dissect many biological processes at the molecular level. Most spectacular is the technology of molecular biology that allows identification and characterization of genes that participate in the genesis of human cancers. Three major groups of genes appear to play out the drama of cancer development: tumor suppressor genes, mismatch repair genes, and oncogenes. The tumor suppressor genes encode products that are inhibitory to cell proliferation.[1] The loss of these inhibitors, by mutation or deletion, can unleash cells from their restraints to proliferate. Mutations in the mismatch repair genes also have been implicated in tumorigenesis.[2-10] The inability of cells to repair spontaneously occurring mutations leads to genomic instability and could potentially result in the accumulation of carcinogenic DNA lesions. Finally, activation of proto-oncogenes, which are normal cellular genes, into oncogenes could accelerate the processes of cell proliferation.[11] c-*myc* was one of the very first proto-oncogenes to be identified and because it normally plays pivotal roles in both cell proliferation and cell death has enticed many to investigate the molecular mechanisms by which it transforms cells.

This monograph is a distillation of many brilliant contributions to the understanding of c-*myc* function in neoplasia from astute observers over the past several decades. Herein, the gene will be denoted in italics (*myc*) and the protein referred to with a capital M (Myc). Given that 6000 papers in which *myc* is a keyword have been published since 1976, we will limit our review of articles which are directly relevant to c-Myc function as well as to some which touch upon the role of *myc* in human cancers. We apologize for omissions of original contributions, specifically when review articles are cited. This introductory chapter provides background information on the regulation of cell proliferation, differentiation, and programmed cell death. The second chapter is an introduction to the basic concepts of tumor suppressor genes, mismatch repair genes, and oncogenes. Subsequent chapters addressing the c-*myc* gene and its product c-Myc, are organized in a fashion to reflect our understanding of the gene, its transcription into messenger RNA, and the translation of mRNA into the different Myc polypeptides. The function of the c-Myc protein will be covered in the remaining several chapters.

1.2 REGULATION OF CELL PROLIFERATION

The body is composed of differentiated cells that perform highly specialized functions, but it also requires accessibility to cells that retain the ability to proliferate, particularly in tissues in which high cell turnover occurs. The balance between cell proliferation and differentiation must be tightly regulated. As an example, the average adult man or woman produces 200 to 300 kilograms of blood cells each year to maintain the steady-state numbers of circulating blood cells. Such high cellular turnover suggests that the control of hematopoietic cell proliferation and differentiation must be exquisitely controlled, otherwise a slight increase in proliferation rate could result in leukemia. A tip in the balance in the opposite direction could lead to severe cytopenias.

How does a cell that retains the ability to proliferate "decide" to leave the resting state and enter the cell cycle? The developmental programs for each mature cell originate at the cell surface where cell-extracellular matrix, cell-cell, and cell-growth factor/lymphokine interactions are translated into intracellular signals and conveyed to the nucleus where changes in gene expression occur.

1.2.1 Signal Transduction: A Maze of Molecular Interactions

The best studied interactions are those between cells and growth factors, which are glycopolypeptides that are secreted by distant (endocrine) or neighboring (juxtacrine) cells, or by the recipient cell (autocrine) in an auto-stimulatory fashion. The receptors for growth factors are transmembrane proteins composed of extracellular, intramembrane, and cytoplasmic domains. Some receptors have intrinsic protein kinase activities while others lack cytoplasmic kinase domains. Some of the latter require interactions with membrane associated G-proteins.

Many of the salient features of growth factor receptor function are best illustrated by the platelet-derived growth factor receptor (PDGF-R), epidermal growth factor receptor (EGF-R), and the interleukin-2 receptor (IL-2R) (Fig. 1.1). When PDGF binds to its cognate receptor, the intrinsic tyrosine kinase activity of the PDGF receptor becomes activated resulting in autophosphorylation of PDGF-R and phosphorylation of specific cytoplasmic proteins.[12-14] Phosphotyrosines present on target proteins enable them to bind non-covalently to SH2 (Src-homology region 2) domains of other proteins. For example, phosphotyrosines on PDGF-R bind the SH2 domain of the γ1 isoform of phospholipase

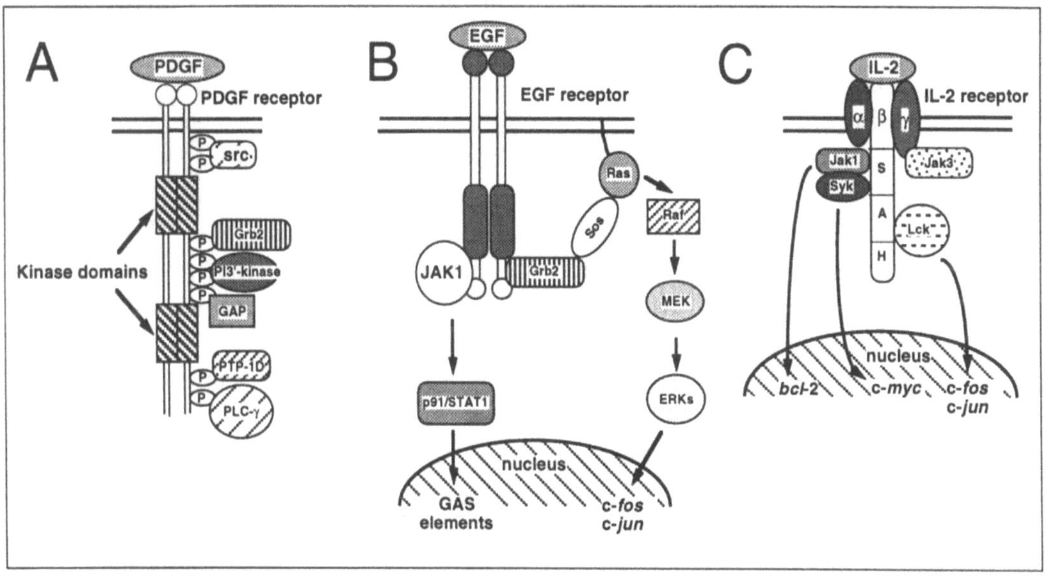

Fig. 1.1. Mitogenic signal transduction through receptor tyrosine kinases. A) The platelet-derived growth factor receptor (PDGF-R) is shown to depict the myriad of interactions between the cytoplasmic tail of PDGF-R and signal transduction proteins. After binding to its ligand, PDGF, the PDGF-R autophosphorylates tyrosine residues (P) that recruit proteins containing SH2 domains: src, Grb2, PI3' kinase, GAP, protein tyrosine phosphatase 1D (PTP-1D), and phospholipase Cγ. B) The epidermal growth factor receptor (EGF-R) is shown to activate the JAK (Janus type kinase)-STAT (transcription factor) pathway as well as the Ras-MAPK (or ERK) pathway. Grb2 engages the activated and phosphorylated EGF-R and recruits Sos, which interacts with Ras initiating the MAPK pathway. C) The IL-2 receptor illustrates the activation of different genetic programs from a single receptor. IL-2 activates the tripartite IL-2R resulting in receptor autophosphorylation. Recruitment of different intermediary messengers to the phosphorylated IL-2R results in the activation of separate pathways. The combination of any two pathways appears sufficient for mitogenesis.

C (PLC γ1). This association results in the phosphorylation of PLC γ1 which augments its enzymatic activity. Activation of PLC γ1 catalyzes the conversion of phosphatidylinositol to diacylglycerol and inositol (1,4,5) trisphosphate, which both act as second messengers in the relay of signals from the cell plasma membrane.[15] Diacylglyerol activates protein kinase C (PKC), which in turn phosphorylates specific cytoplasmic proteins. Inositol trisphosphate induces the release of intracellular calcium, which activates calcium-dependent protein kinases.[16] The waves of protein phosphorylations originating from various points within the cytoplasm converge on the cell nucleus to elicit responses through alterations in gene transcription.

Several features of the EGF-R distinct from those of PDGF-R emphasize other noteworthy concepts about signal transduction.[17] The binding of EGF to EGF-R leads to autophosphorylation of tyrosine residues of the EGF-R. The phosphotyrosines then allow EGF-R to bind the SH2 domain of Grb2 which exists as a cytoplasmic heterodimeric complex with Sos. Sos binds to Ras in this quaternary complex with the consequent activation of Sos as a guanine nucleotide exchange protein and conversion of Ras from the inactive Ras-GDP form to the activated Ras-GTP form. Ras-GTP, in turn, could associate with and activate the Raf-1 serine-threonine protein kinase. Raf-1 is known to activate MAP kinase kinase, which itself activates the mitogen-activated protein (MAP) kinase. MAPK, in turn, phosphorylates a number of transcription factors including the interferon activating factors (STAT), Myc, and Fos with subsequent alterations in their transcriptional properties.

The interleukin-2 receptor (IL-2R) illustrates the participation of the nonreceptor Src-like protein tyrosine kinases in signal transduction.[18-20] Nonreceptor tyrosine kinases such as Lck and Syk have been linked to specific cell surface receptors. Lck, for example, is associated with CD4, a receptor made famous by its association with the deadly human immunodeficiency virus. Both Lck and Syk have been shown to participate in the signalling of the IL-2R that is composed of three subunits: α, β and γ. Upon engagement with IL-2, separate regions of the cytoplasmic tail of the receptor β subunit become tyrosine phosphorylated. These phosphorylated domains recruit Syk and Lck, which in turn induce the expression of c-*myc* and c-*jun*/c-*fos*, respectively. Thus, the nonreceptor tyrosine kinases that contain SH2 domains allow them to be recruited to specific phosphotyrosine containing proteins to engage in delivering messages from the extracellular world to the nucleus.

1.2.2 Transcription Factors: the Final Players

Transcription regulatory factors are considered to be the final players in the afferent signalling pathway from the cell plasma membrane.[21-24] Transcription factors are proteins that bind specific cognate DNA sequences, either directly or indirectly, to alter the rate of transcription through protein-protein interactions with the basal transcriptional machinery (Fig. 1.2). Many genes contain a DNA sequence called a TATA-box that allows the assembly of the basal transcriptional machinery, which is composed of several general transcription factors such as TFIID. TFIID, whose master subunit is the TATA-binding protein (TBP). Binding of TFIID to the TATA box is facilitated by another general transcription factor called TFIIA and is followed by the sequential binding of other general transcription factors and the recruitment of RNA polymerase II, the enzyme required for transcription of messenger RNA. The triggering of RNA polymerase II to synthesize heterogenous nuclear RNA appears to involve phosphorylation of its highly repeated structure tail, termed CTD (carboxy-terminal domain), by the basal factor TFIIH. TFIIH is intriguing, since it is a multisubunit enzyme consisting of several polypeptides including a protein kinase that is virtually identical to the CDK-activating kinase (CAK).[25,26] How transcription and cell cycle regulation is linked through TFIIH is currently under investigation.

Regulation of eukaryotic gene expression requires additional proteins such as sequence-specific transcription factors and accessory factors called TBP-associated factors (TAFs), that augment or diminish the activity of the basal transcriptional machinery.[22,23] Through protein-protein contact, transcription factors and TAFs interact with the RNA polymerase II-basal transcriptional complex to regulate transcription of specific genes. Although these interactions have traditionally been thought to result in activation of gene expression, certain transcriptional activators in the context of TAFs, RNA polymerase II, and perhaps other proteins can actually repress transcription. Transcriptional repression will be discussed in more detail in chapter 10.

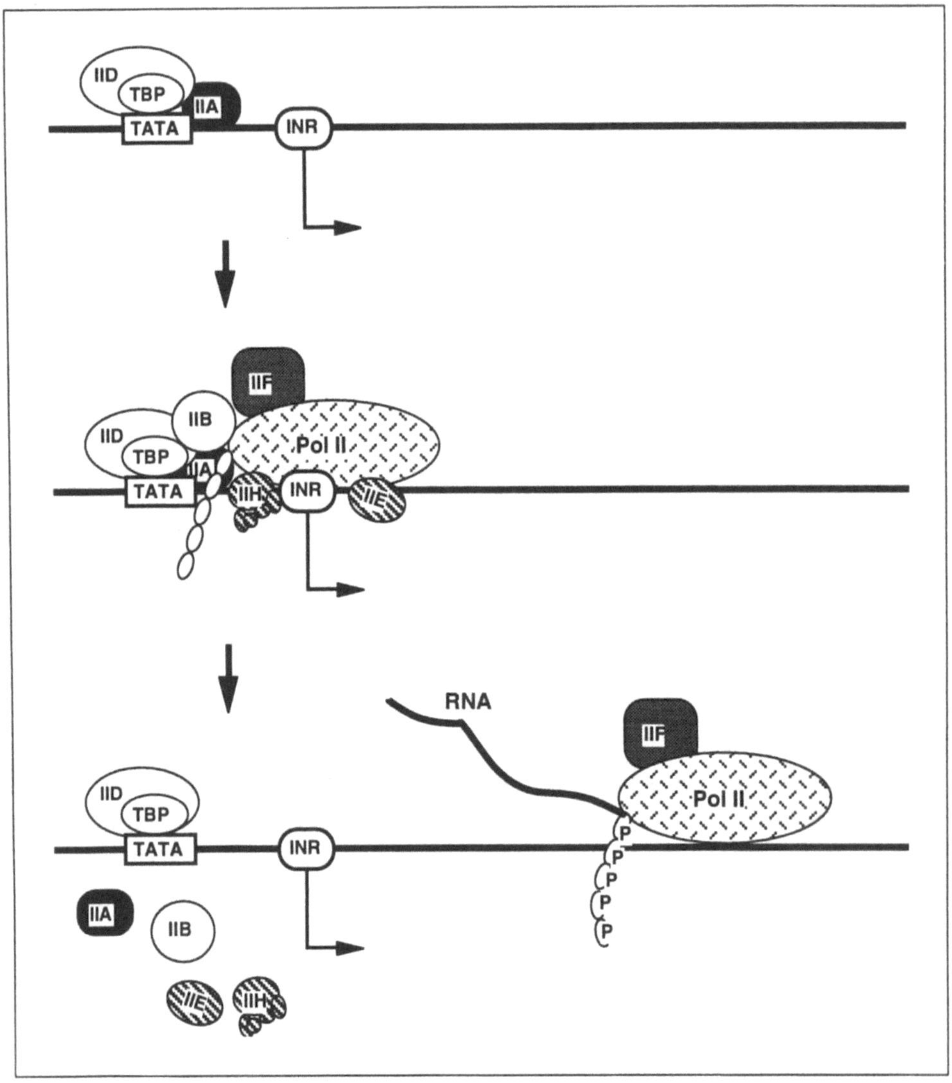

Fig. 1.2. The basal transcriptional machinery. The top diagram depicts a minimal promoter containing a TATA and initiator (INR) element. TATA binding protein (TBP) is shown tethered to the TATA element as part of the TFIID complex. Subsequent recruitment of basal transcription factors (labelled II followed by a letter) culminates in the transcriptional initiation complex (middle diagram). The bottom diagram depicts the transcribing of heterogenous nuclear RNA associated with phosphorylation of the RNA polymerase IIA subunit tail (small repeat open ovals). TFIIH is a multisubunit factor, which has cyclin dependent kinase and helicase activities.

Transcription factors are composed of distinct modular functional domains that confer specific activities to these proteins such as DNA-binding, dimerization, and transcriptional activation (Fig. 1.3A).[27] Transcription factors contain transcriptional activation domains which do not have rigid structural motifs, although several classes of activation domains have been recognized by their amino acid content. Prototypical activation domains are the acidic domain of *Herpes simplex* viral VP16 protein, the glutamine-rich domain of Sp1, and the proline-rich domain of NF1. The activation domains are thought to bind to specific co-activator proteins, such as TAFs that are components of the macromolecular transcriptional machinery. Presumably allosteric alterations link the activation domains to TFIID and RNA polymerase II via the co-activators.

Various classes of specific DNA-binding domain motifs have been identified. Most DNA binding domains are comprised of helical regions that bind specifically in the major groove of DNA; other secondary structures that participate in specific DNA binding have also been identified. Alignment of the DNA binding regions is achieved by tertiary protein structure or by regions of the transcription factor that mediate homodimerization or heterodimerization with other proteins. Combinatorial associations between transcription factors further expand the repertoire of binding sites for transcriptional regulation.

Transcription factors are classified according to their structural differences that include the leucine zipper, helix-loop-helix, helix-loop-helix-zipper, helix-turn-helix, and zinc finger motifs (Fig. 1.3B).[28] The leucine zipper (Zip) is a slightly bent α-helical coil composed of a region of basic amino acids and a motif in which every seventh residue is a leucine.[29] The Zip interacts with the corresponding Zip of a partner protein which aligns the basic regions in a "scissor-like" configuration to interact with specific bases in the major groove of DNA. Typical members of this family are the Fos and Jun proteins, which can form either homodimers or heterodimers with proteins of the same transcription factor family. The helix-loop-helix (HLH) family of proteins includes the prototypical myogenic protein MyoD and the ubiquitous factor, E2A.[30] The HLH motif contains two α-helical regions separated by a DNA loop. Helix-loop-helix proteins resemble leucine zipper proteins in that members also dimerize to form homodimers or heterodimers. A separate class of transcription factors, of which Myc is a member, contains both the HLH and Zip motifs. Dimerization is mediated by the HLH and Zip, which together with helix 2 of the HLH form a long continuous helical domain. Myc heterodimerizes to another HLHZip protein called Max, but does not appear to form homodimers.

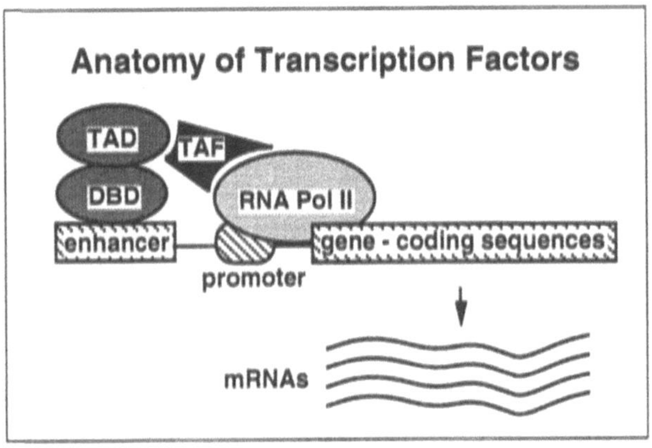

Fig. 1.3.A. Schematic depiction of a generic transcription factor. Transcription factors are generally composed of a DNA binding domain (DBD) and a transcriptional activation domain (TAD). Transcriptional repressors have domains that actively suppress transcription (not shown). The transactivation domain influences the basal transcriptional machinery through protein-protein interactions via co-activators or TAFs (TBP associated factors) that in turn regulate the basal transcriptional complex including RNA polymerase II.

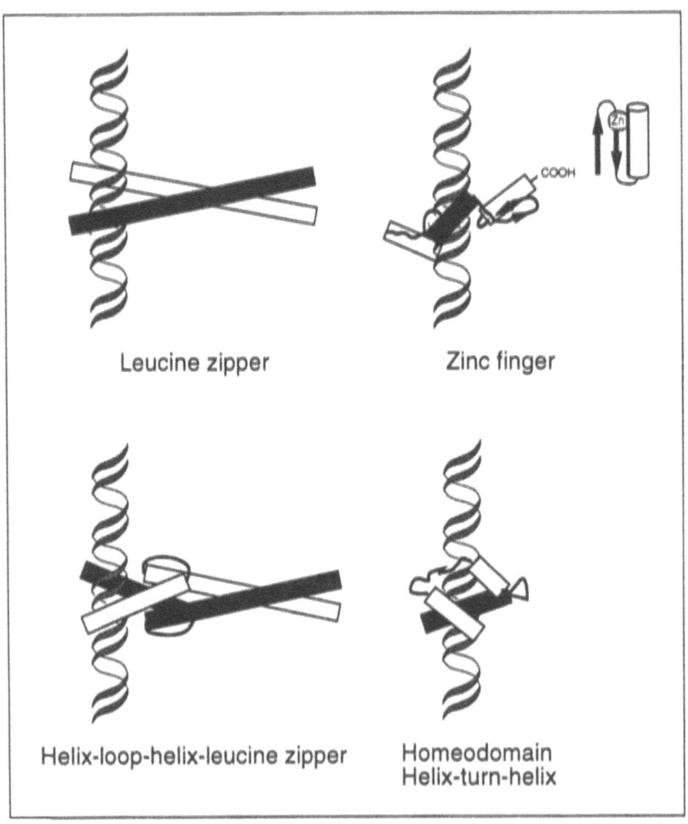

Fig. 1.3.B. Transcription factors have been classified according to their DNA binding motifs. Several of these motifs are shown. The bars represent helical protein domains. The zinc finger motif is depicted separately (top right) as a helix contiguous with turns of β sheets (thick arrows). Note that most DNA binding motifs contain helical protein basic domains that interact with nucleotide bases in the major groove of DNA. Not shown are other DNA binding motifs which include minor groove binding proteins or proteins that interact with DNA through non-helical structures.

1.2.3 The Cell Cycle in Growth Regulation

The proliferating cell and its descendants may be viewed as travelers through a cyclic series of cellular states, collectively termed the cell cycle (Fig. 1.4).[31-34] Although the cell cycle is ultimately controlled by extracellular signals, familiarity with the intrinsic components of the cell cycle are vital to understanding the regulation of cell growth and cancer. The cell cycle is composed of phases identified by unique morphological and biochemical characteristics of cells as they traverse this reproductive cycle to produce progeny cells. The normal cell cycle is marked by two distinct phases, S (DNA synthetic phase) and M (mitosis), that are separated by two gaps, G_1 that precedes S phase, and G_2 that follows S phase. Cells in the resting state are said to reside in G_0. The proteins that act as gatekeepers, checks, and balances to regulate the orderly progression of cells through the cell cycle are beginning to

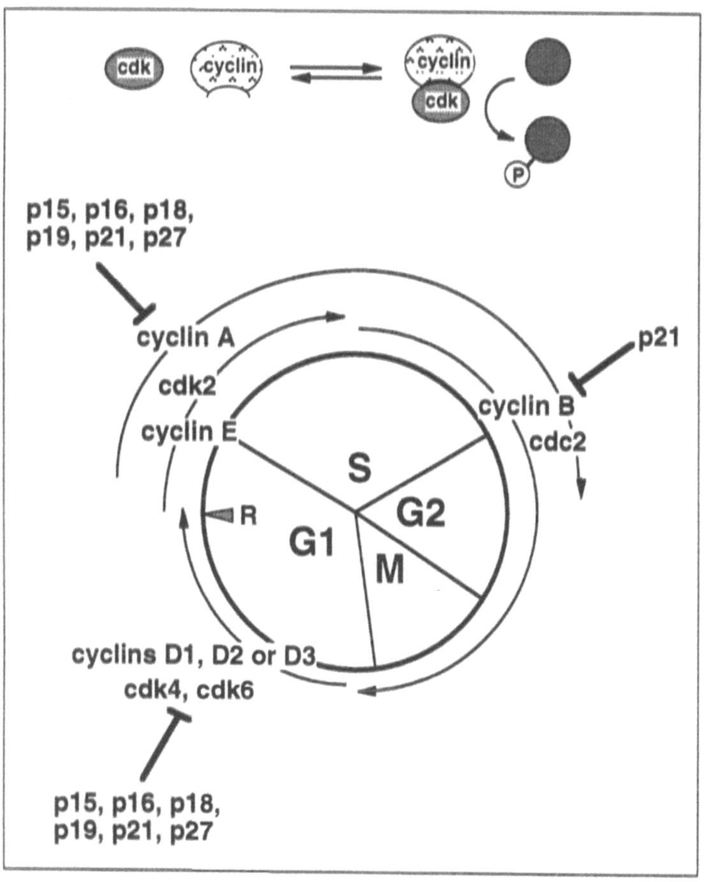

Fig. 1.4. The cell cycle. The cell cycle is regulated by protein kinases consisting of a catalytic subunit or cyclin dependent kinase (CDK) and a regulatory subunit or cyclin (top diagram). The activated cyclin-CDK complex is shown phosphorylating a substrate (shaded circle) at Ser or Thr. The bottom diagram depicts the different phases of the cell cycle: G1, S, G2 and M. The cyclins or CDKs involved in regulating the different phases of the cell cycle are depicted. Inhibitors of cyclin dependent kinases (p15, p16, p18, p19, p21, p27) are shown with their sites of action. Within the G1 phase, R represents the restriction point beyond which cells are commited to enter S phase.

emerge as participants in a cyclical dance. In fact, the normal cell cycle may be viewed as the result of coupled molecular harmonics (oscillations of protein activities) linked in series and in parallel. In cell culture models, when nutrients are omitted the cells withdraw from this cyclical dance into G_0.

The major gatekeepers that regulate the orderly procession of cells through the cell cycle consist of regulatory subunits, termed cyclins, and their associated catalytic serine-threonine kinases called cyclin-dependent kinases (CDKs). Cyclins are proteins whose levels oscillate through the cell cycle, while levels of CDKs are generally constitutively present. The specific assembly of cyclins with their cognate CDKs into active complexes results in oscillatory kinase activities with substrates such as the retinoblastoma protein, pRb, and c-Myc. The activity of the cyclin-CDK complexes is countered by the CDK inhibitor (CDI) proteins. These proteins appear to be responsible for growth suppression by inactivating the cyclin-CDK complexes at different points in the cell cycle. The complex levels of regulation by these proteins will be discussed in the following sections.

1.2.3.1 G1 Phase and Its Molecular Gatekeepers

G_1 is the gap following mitosis and preceding S phase. The duration of G_1 is highly variable among different cell types and is regulated by the availability of nutrients and growth factors. Cells grow in mass during G_1 to a certain size and reach a critical checkpoint called the restriction (R) point, at which time the cell becomes committed to DNA synthesis (Fig. 1.4). Serum or growth factors are no longer needed after the R point for the cells to enter into S-phase. TGFβ could no longer inhibit the growth of certain cells after the restriction point.[35] Thus the R point represents a critical juncture for the regulation of cell proliferation. This juncture is regulated by a series of reactions involving G_1 cyclins and pRb.

The principal regulators of G_1 progression are cyclins D and E and their associated kinases, cdk2, 4, 5 and 6.[36,37] The D-type cyclins are mitogenically induced in a cell-lineage specific manner as delayed early response genes.[38] Introduction of antibodies to D-type cyclins (of which there are three, D_1, D_2, D_3) into cells results in failure of the cells to enter S-phase, while overexpression of cyclin D in nontransformed cells accelerates their entry into S phase confirming the critical role of the D-type cyclins in cell cycle progression.[39] Both the CDK proteins and the D-type cyclins exert their positive control on the cell cycle presumably by phosphorylating and inactivating negative regulators of the cell cycle, like pRb and the pRb-related protein p107.[40,41] pRb is thought to arrest cells in G1 by binding to transcription factors such as E2F, which are necessary for activating genes required for progression into S phase.[42-46] In support of this mechanism of pRb inactivation, hypophosphorylated pRb has been found in differentiated cells and in cells exposed to differentiating agents. Exposure of cells to growth factors leads to an increase in the expression of the D-type cyclins and cdk4 and ultimately results in the phosphorylation of pRb by cyclin D1-cdk4 (Fig. 1.5).

Phosphorylation of pRb disrupts its association with E2F, allowing cells to proceed into S phase. It has been alternatively postulated that the pRb-E2F complex actively represses transcription, and phosphorylation of Rb disrupts its interaction with transcriptional suppressing proteins, resulting in a loss in transrepression. E2F released from its restraints by pRb is able to activate the expression of cyclin E. Cyclin E levels peak in late G_1 and its overexpression causes a shortened G_1 duration presumably due to early entry into S.[47-50] Once cells enter S phase, cyclin E is degraded, increasing the availability of cdk2 for binding to cyclin A, a key regulator of the transition from S to G2 phases.

The transition through G_1 phase is partly regulated by the CDI proteins. G_1 arrest occurs after exposing cells to ionizing radiation by inducing the expression of a protein called p53.[51] G_1 arrest allows time for the detection of DNA damage and DNA repair prior to the induction of DNA synthesis. p53 causes G_1 arrest by inducing the expression of p21, a member of the CDK inhibitor (CDI) family.[52] The CDI proteins, which now include, p15, p16, p18, p19, p21, and p27, bind and inactivate cyclin-CDK directly. p21 is transcriptionally regulated by p53 and interacts with a variety of CDK complexes containing cyclins A, B, D, and E. p27, which also inhibits CDK complexes, is involved in the response to external signals like the growth factor TGFβ that arrests cells in G1 by interfering with cyclin E-cdk2 mediated pRb phosphorylation. TGFβ probably increases the availability of p27 to associate with cyclin E-cdk2 complexes and prevent cdk2 activation since p27 is expressed constitutively even in the absence of TGFβ.[53,54]

G_1 arrest induced by cell contact inhibition also appears to be mediated through p27. Unlike p21 or p27, the homologous proteins p16 and p15 appear to specifically inactivate cyclin D-cdk4 complexes. The critical role of CDIs in cell cycle regulation is emphasized by the recent discovery that a tumor suppressor gene encodes p16.[55] Mutations of p16 have been found to occur in several human tumors and a germline mutation in a family with fa-

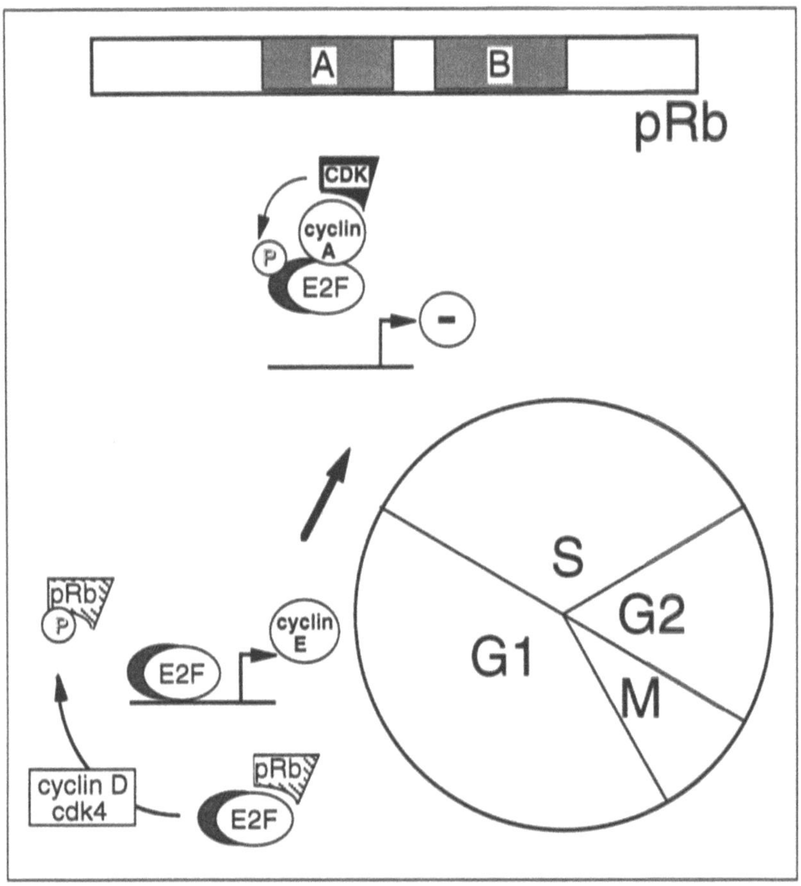

Fig. 1.5. The retinoblastoma protein (pRb) and its regulation of the cell cycle. pRb is depicted at the top as a bar containing the pocket domains, A and B, which interact with specific cellular proteins as well as viral proteins like SV40 large T antigen. pRb binds the cellular transcription factor E2F and inhibits its function. Cyclin D-cdk4 phosphorylates pRb late in the G_1 phase to release active E2F, which is depicted activating the expression of cyclin E. Cyclin E protein forms a complex with cdk2, which further phosphorylates pRb in a feed-forward loop and drives the cell into S phase. Cyclin A, which is presumably induced by E2F and cyclin E, is required for DNA synthesis. Cyclin A-cdk2 participates in a negative feedback loop by phosphorylating the dimerization partner of E2F, DP1 (shaded oval), to inhibit DNA binding.

milial melanoma has been identified.[56] It has been questioned, however, whether p16 mutations are acquired in cultured tumor cells and do not exist in fresh tumor explants.[57,58]

1.2.3.2 Entry into S and Beyond

The entry of a normal cell into S phase is preceded by proper cellular growth and DNA damage assessment and repair. Prior to or upon entry into S, cyclin A levels rapidly increase.[46,59-64] Cyclin A appears to complex with cdk2, as well as to proliferating cell nuclear antigen which is a subunit of DNA polymerase.[63] This association suggests a possible direct link between the cyclins and the DNA replication machinery. Although we are gaining a better understanding of the mechanisms of certain gatekeepers, many unsolved mysteries abound. For example, how the cell senses the completion of DNA replication and inhibits the re-initiation of DNA replication remains an enigma.[32] It is known that cyclin B levels begin to increase in late S-phase and perhaps its presence somehow prevents re-initiation of DNA replication.

The entry of the cell into G_2 is associated with the persistence of cyclins A and B.[65,66] Cyclin A levels appear to diminish in late G_2, whereas cyclin B levels persist until mitosis. Cyclin B associates with the prototypical CDK, $p34^{cdc2}$, to form a holoenzyme complex that is known to be the major regulator of mitosis.[67-69] The cyclin B kinase complex phosphorylates nuclear lamins and causes the disassembly of the nuclear envelope prior to the initiation of chromosomal segregation.

Upon completion of cell division or cytokinesis, the daughter cells restart in G_1 the cyclical dance among the families of CDK proteins and cyclins. If the cell senses an inadequacy of nutrients, growth arrest ensues and the cell withdraws in the resting phase G_0. Even with appropriate nutrients, normal tissues, such as the liver, restrain cells in G_0, until loss of cells due to injury necessitates recruitment of resting cells into the cell cycle to fill the void.

1.3 CONTROL OF CELL DIFFERENTIATION

Cells in vertebrate animals are destined to proliferate, differentiate or die. Diverse cellular functions required for the development, survival and propagation of animals are ensured by differentiation of cells during embryogenesis. Differentiation of cells from progenitors continues to occur in adulthood as old and dead cells are replaced by new ones. The understanding of cell differentiation remains embryonic; however, a number of models have provided the conceptual framework for mechanisms controlling cellular differentiation. Here we review several experimental systems that have revealed some of the molecular switches involved in regulating differentiation.

1.3.1 ADIPOGENESIS

The in vitro tissue culture model for differentiation of 3T3-L1 fibroblastic preadipocytes into adipocytes exemplifies the interaction between extracellular factors and nuclear transcription factors in differentiation.[70] In this model, confluent 3T3-L1 cells are induced to differentiate by a cocktail of factors such as insulin and dexamethasone. The cells then undergo several rounds of division, cease to proliferate, and then acquire the biochemical and morphological phenotype of differentiated adipocytes. Differentiation of cultured adipocytes is accompanied by the regulated expression of adipose-specific genes. Characterization of the *cis*-acting regulatory elements within the promoters of adipose-specific genes has revealed sequences that are responsible for activating or derepressing transcription during differentiation.

The transcription factor C/EBPα has been extensively studied among *trans*-acting factors that participate in the expression of adipose-specific genes. C/EBPα is the prototypical leucine zipper protein that binds the CAAT box motif found in promoters of adipose-specific genes. The expression of C/EBPα follows the induction of adipocyte differentiation and precedes the transcription of adipose-specific genes. Experiments using ectopic antisense or enforced sense C/EBPα expression suggest that C/EBPα is necessary and sufficient for differentiation of 3T3-L1 into adipocytes.[71] Other members of the C/EBP family of proteins are also expressed at the time of differentiation although their levels decline thereafter. It is noteworthy that regulatory sequences far removed (in terms of DNA base pairs) from promoters of adipose-specific genes are also required for tissue-specific expression. Thus, both proximal and distal regulatory sequences may be required for many tissue-specific gene expressions.

The mechanisms linking extracellular differentiation-inducing agents to nuclear events is beginning to be revealed through the study of promoters of genes that respond to extracellular signals. The promoter region of the *c/ebp* α gene, for example, contains an array of putative transcription factor binding sites including those of c-Myc/USF, Krox 24, CUP, Spl and C/EBPα itself. The c-Myc binding site is within a potential repressing *cis* element within the *c/ebp* α promoter. Enforced expression of c-Myc in 3T3-L1 cells prevents the expression of C/EBPα and differentiation into adipocytes.[72] The c-*myc* gene expression pattern is an approximate mirror-image of the expression of *c/ebp* α during adipocyte differentiation suggesting that release from suppression by silencing *cis*-elements may be a critical initiating event.

1.3.2 MYOGENESIS

The cell culture model of myogenesis has provided astounding molecular insights into the mechanisms of how cells differentiate (Fig. 1.6).[73-76] Early studies of myogenesis suggested that a single master gene may control muscle cell differentiation.[77,78] Treatment of C3H $10T^1/_2$ fibroblasts with 5-azacytidine induced differentiation resulting in many myogenic colonies and fewer numbers of chondrogenic and adipogenic colonies. Transfection of genomic DNA from 5-azacytidine-treated myoblasts, but not DNA from parental cells, converted C3H $10T^1/_2$ cells into myocytes suggesting that some myogenic gene was activated by 5-azacytidine.[77] The responsible gene was cloned and termed *myoD*.[78] It is specifically expressed in myoblasts and not in the parental C3H $10T^1/_2$ cells, and its sole expression is sufficient to convert C3H $10T^1/_2$ cells into myocytes. Overexpression of c-*myc* is able to inhibit myogenic differentiation suggesting that signals for cell proliferation could override cues for differentiation.

MyoD is a member of the HLH family of myogenic transcription regulators that bind to consensus DNA sites (CANNTG or E-box sequence) present in most muscle-specific enhancers.[74,75,79,80] Other family members such as myogenin, myf-5, and myf6-mrf4-herculin can also convert C3H $10T^1/_2$ cells into myocytes. The MyoD protein contains an N-terminal transcriptional activation domain and a basic-helix-loop-helix (bHLH) that heterodimerizes with the ubiquitous bHLH E2A transcription factors to bind DNA..[81] The association of MyoD with E2A is regulated by a third factor, Id, that contains an HLH motif but lacks the DNA-binding basic region (Fig. 1.6).[82,83] Thus, heterodimers of Id and MyoD or E2A proteins are unable to bind DNA. Ectopic expression of Id is able to inhibit the activity of MyoD and retard myogenic differentiation. The myogenic differentiation model has led to the discovery of a network of protein-protein interactions involving factors, such as MyoD and Id, that regulate muscle-specific gene expression. The paradigms generated by these studies have provided a basis for the understanding of other programs of cell differentiation.

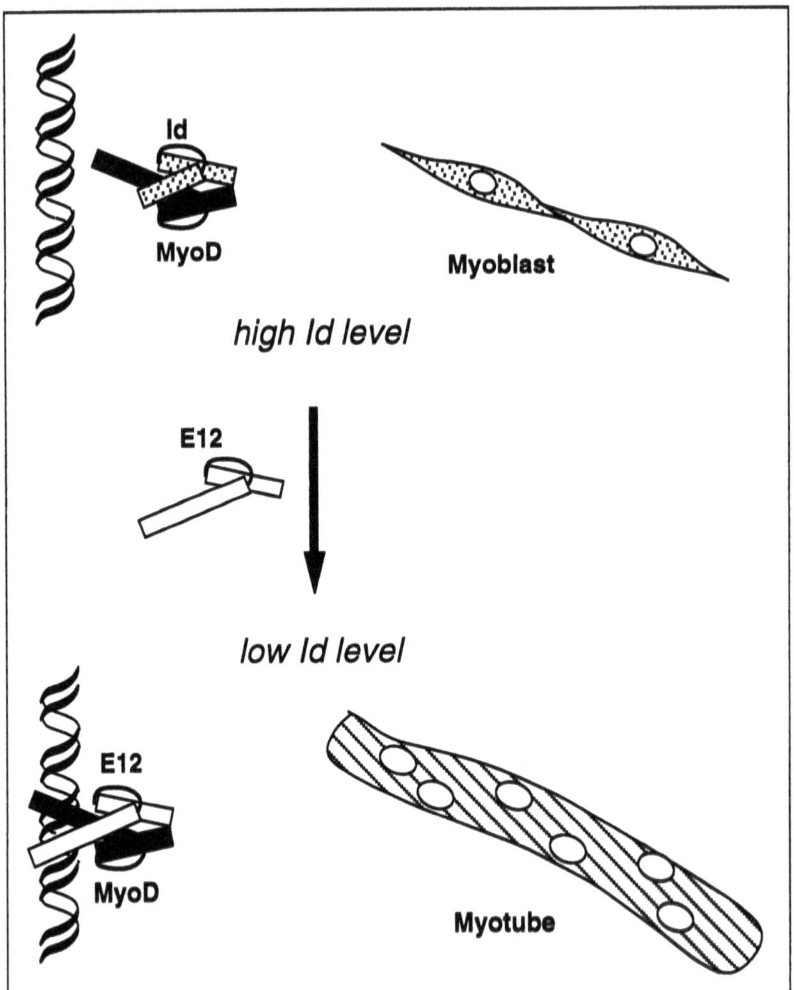

Fig. 1.6. Helix-loop-helix proteins regulate myogenesis. The figure depicts the differentiation of myoblasts into a myotube. Id, an HLH protein lacking a basic region, forms inactive heterodimers with MyoD in myoblasts. With differentiation, the level of Id decreases and E12 forms DNA binding heterodimers with MyoD that activate the myogenic program.

1.3.3 HEMATOPOIESIS

Hematopoiesis is the only differentiation system that is amenable to extensive ex vivo studies.[84] Pluripotent hematopoietic stem cells (PHSCs) isolated from bone marrow differentiate into many highly specialized circulating blood cells. The long-term repopulating PHSCs self-reproduce and also differentiate toward the common lymphoid or myeloid multipotent stem cell. PHSCs are dormant unless stochastically awakened to enter the cell cycle. The lymphoid stem cell differentiates into either mature T or B lymphocytes. The myeloid multipotent stem cell differentiates into a variety of circulating blood cells including basophils, eosinophils, neutrophils, monocytes, red blood cells, and platelets. The developmental programs for each mature circulating cell consist of networks of cell-extracellular matrix, cell-cell, and cell-growth factor/lymphokine interactions.

Commitment of stem cells to differentiate is a stochastic process. In pair daughter cell studies, daughter cells derived from a single parent progenitor cell frequently spontaneously differentiate into different cell lineages. Such observations suggest that random activation of groups of differentiation genes might underlie lineage specific differentiation. How then are the specific genetic programs generated and selected for the different lineages of cells? Selection of randomly generated genetic programs is associated with the known random shuffling of immunoglobulin gene fragments through V(D)J rearrangements to generate antibody diversity. Clones of B-cells are then selected by antigens that specifically engage antibody variable regions arising from random V(D)J rearrangements. The stochastic differentiation of hematopoietic cells may in principle be similar to that for generation of antibody diversity. With hematopoiesis, different hematopoietic growth factor receptors may be stochastically expressed and cells are selected for survival by cell-cell contact as well as growth factors.

Lineage-specific factors support the survival, proliferation, and maturation of progenitors that are committed through a hypothetical stochastic expression of specific groups of differentiation genes. For example, macrophage colony-stimulating factor (M-CSF) promotes the proliferation and differentiation of macrophage/monocytes. Erythropoietin promotes the survival, proliferation, and differentiation of erythrocytes. IL-5 and granulocyte colony stimulating factor (G-CSF) are thought to be specific for eosinophils and granulocytes, respectively. A late acting lineage-specific factor for megakaryocytopoiesis has been identified and termed thrombopoietin or megakaryocyte differentiation and growth factor. These late-acting lineage-specific factors could be viewed as factors that select out and promote the proliferation of a subset of committed cells whose differentiation programs have been randomly selected. Lineage specific transcription factors, such as GATA-1, constitute a part of the differentiation program.[85-89] The activation of groups of differentiation genes that are not supported by lineage-specific factors hypothetically could result in cell death.

1.4 CONTROL OF CELL DEATH

Apoptosis is a unique process of cell death that is morphologically distinct from necrosis (Fig. 1.7).[90-93] Necrosis is accidental cell death resulting from severe and sudden thermal, physical, or chemical trauma that cause early mitochondrial and cellular swelling with ensuing cytoskeletal disruption and ruptured plasma membrane and organelles. The nucleus remains essentially intact throughout necrosis. In contrast, apoptosis was initially recognized as a process starting with condensation of the chromatin and shrinkage of cell volume.[94,95] The plasma membrane then becomes ruffled and blebbed. The nucleus and cytoplasm finally become partitioned into membrane-bound apoptotic bodies that are shed from the dying cells. The term apoptosis was proposed by Kerr, Wyllie and Currie to describe these cellular alterations and was derived from a Greek word that describes the falling of petals from flowers, or leaves from trees.[94] Most cells in the last stages of apoptosis display a characteristic degradation of nuclear DNA into oligonucleosomal sizes of multiples of 180 base pairs (DNA laddering). Throughout apoptosis, the mitochondria remain virtually morphologically normal. Apoptosis participates in many physiological processes including morphogenesis, elimination of self-reactive T-cells, and perhaps death of B-cells with nonproductive V(D)J gene rearrangements.

1.4.1 Genetics of Programmed Cell Death — Lessons from a Round Worm

Although apoptosis is commonly equated with programmed cell death in vertebrate systems, a genetic program for cell death has only been established unequivocally for the death of individual cells

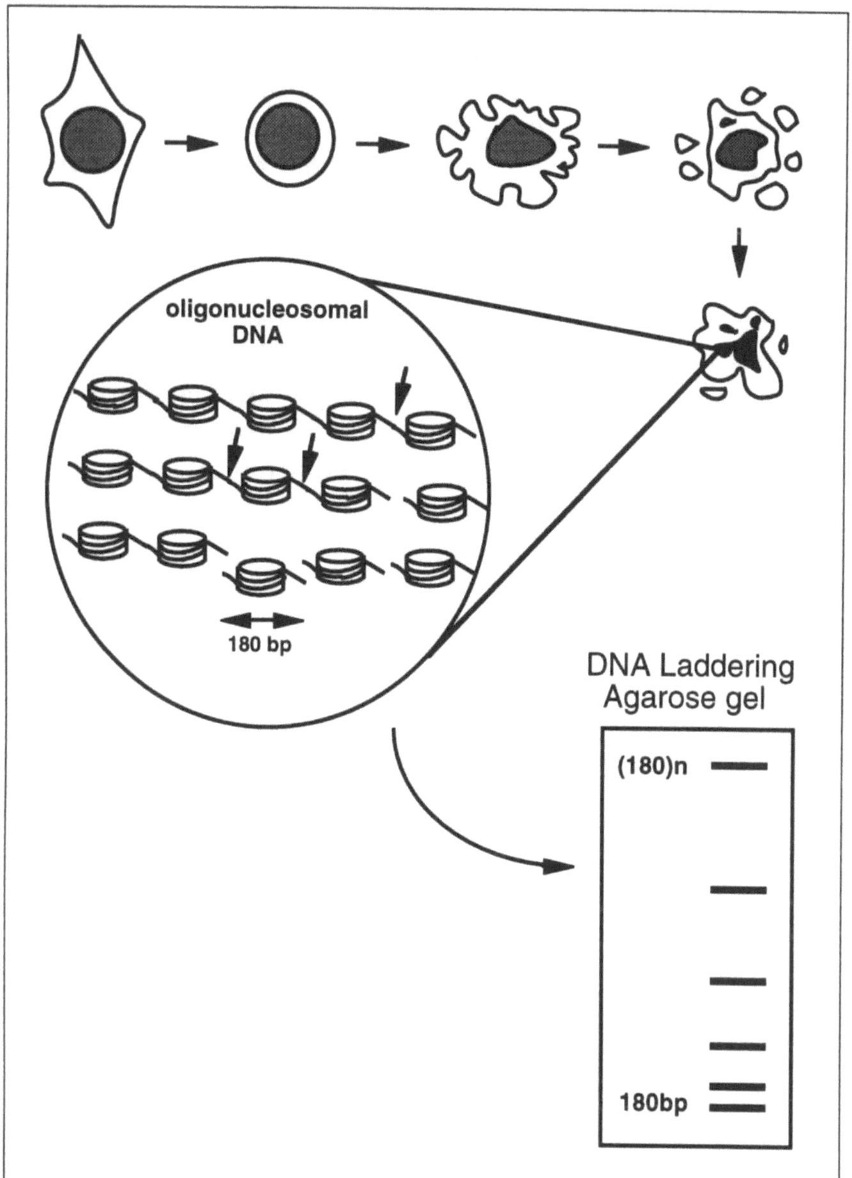

Fig. 1.7. Apoptosis. The term apoptosis was first proposed to describe the morphologic changes associated with certain forms of cellular death. Cells undergo programmed cell death or apoptosis with initial membrane blebbing (top diagram). The blebbing subsequently yields membrane bound cellular debris which are taken up by surrounding cells through phagocytosis. The cell nucleus condenses and the chromatin is fractured by endonucleases that generate oligonucleosomes (middle diagram). Experimentally, DNA extracted from apoptotic cells displays a ladder of DNA fragments with sizes that are multiples of 180 bp, the size of a nucleosome bound piece of DNA.

in the worm, *Caenorhabditis elegans* (Fig. 1.8).[96] During the development of *C. elegans*, 131 out of 1090 somatic cells are programmed to die. Studies of mutant strains of *C. elegans* have revealed at least 14 genes (complementation groups) that are involved in cell death. Of these, *nuc-1* is a gene that encodes an endonuclease which may be involved in the generation of oligonucleosomal DNA fragments during the final stages of apoptosis. Mutant strains containing two genes required for cell death, *ced-3* or *ced-4*, have extra somatic cells.[97,98] The vertebrate gene that is structurally homologous to *ced-3* encodes the interleukin 1β converting enzyme or ICE. ICE is a protease that cleaves pro-interleukin 1β to its active form. Its proteolytic activities may be involved in apoptosis, although recent studies suggest that another protease similar to

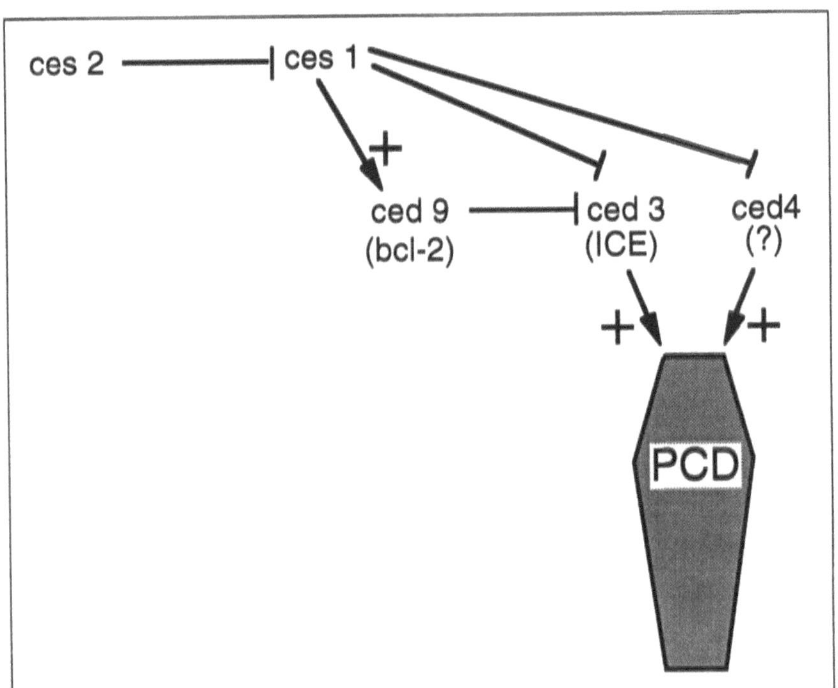

Fig. 1.8. Lessons on apoptosis from the invertebrate Caenorhabtidis elegans. *The worm* C. elegans *develops in a manner in which 131 somatic cells are programmed to undergo apoptosis. Some of the cell death regulatory genes are depicted in this figure.* ces 1 *(helix-loop-helix protein) and* ces 2 *(zinc finger protein) are upstream regulatory genes that influence the expression of* ced 9, ced 3 *and* ced 4. ced 9, *a homolog of the vertebrate* bcl-2, *blocks apoptosis, whereas* ced 3, *an ICE-like protease, is required for programmed cell death (PCD).* ced 4 *is also necessary for apoptosis, although a vertebrate homolog has not been identified to date.*

ICE, termed PRICE, may be responsible for the proteolytic degradation that is associated with the final pathway of apoptosis.[99-101] Mutation of the locus *ced-9*, which antagonizes *ced-3* and *ced-4*, result in developmental arrest. The product of *ced-9* blocks apoptosis and hence its loss presumably results in untimely cell death.[102-104]

1.4.2 Mediators of Apoptosis in Vertebrates

The genetics of programmed cell death in vertebrates is only beginning to emerge. An intriguing genetic clue to vertebrate apoptosis is a characteristic chromosomal translocation [t(14:18)] observed in most follicular lymphomas. This translocation activates a gene found at the breakpoint, *bcl-2*, whose product has significant sequence homology to the *C. elegans ced-9* protein.[105] The expression of Bcl-2 protein is restricted to tissues characterized by apoptotic death. For example, Bcl-2 is found in the thymic medulla where surviving T-cells are located, whereas Bcl-2 is not found in cortical thymocytes destined to die. The function of Bcl-2 at the molecular level still eludes precise biochemical characterization, however the Bcl-2 protein is known to exist in mitochondrial, endoplasmic reticular, as well as perinuclear membranes. Recent evidence suggest that Bcl-2 may alter intracellular Ca^{+2} fluxes or regulate reactive oxygen species.[106,107] Other studies suggest, however, that Bcl-2 does not participate in neutralizing reactive oxygen species.[108] Bcl-2 function is modulated by its interaction with Bax (Fig. 1.9). Bax homodimers tip the balance toward cell death, which may be averted by the formation of Bax-Bcl-2 heterodimers.[107,109-111] In transgenic mice experiments, targeted *bcl-2* expression in B-cells resulted in memory cells with a prolonged life. Targeted *bcl-2* expression in T-cells resulted in immature thymocytes that are resistant to apoptosis induced by glucocorticoids, radiation or anti-CD3 antibody. Thus, follicular lymphomas, which are characterized by a low growth fraction and deregulated expression of *bcl-2*, proliferate predominantly by a decrease in the rate of cell death.

The tumor suppressor protein p53 was recently shown to participate in the regulation cell cycle G_1-S phase transition and apoptosis.(Fig. 1.10).[112-117] It is thought to be involved in the monitoring of DNA breaks induced by various physical and chemical agents such as ionizing radiation. After DNA is broken, p53 protein

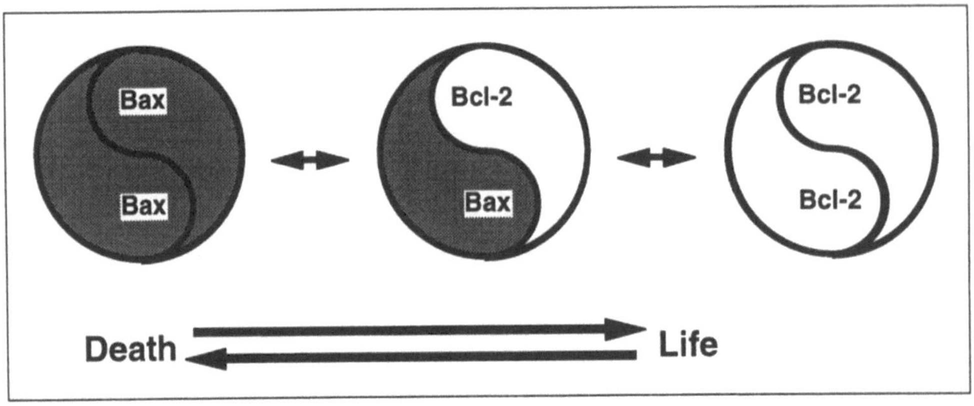

Fig. 1.9. Bcl-2 and Bax family of proteins regulate programmed cell death. Bcl-2 belongs to a large family of related proteins that are capable of homodimerization and specific heterodimerization among the family members. Bax homodimers are thought to favor apoptosis, whereas Bcl-2 neutralizes Bax through heterodimerization. Bcl-2 homodimers are probably inactive, although an active role for these homodimers has not been completely ruled out.

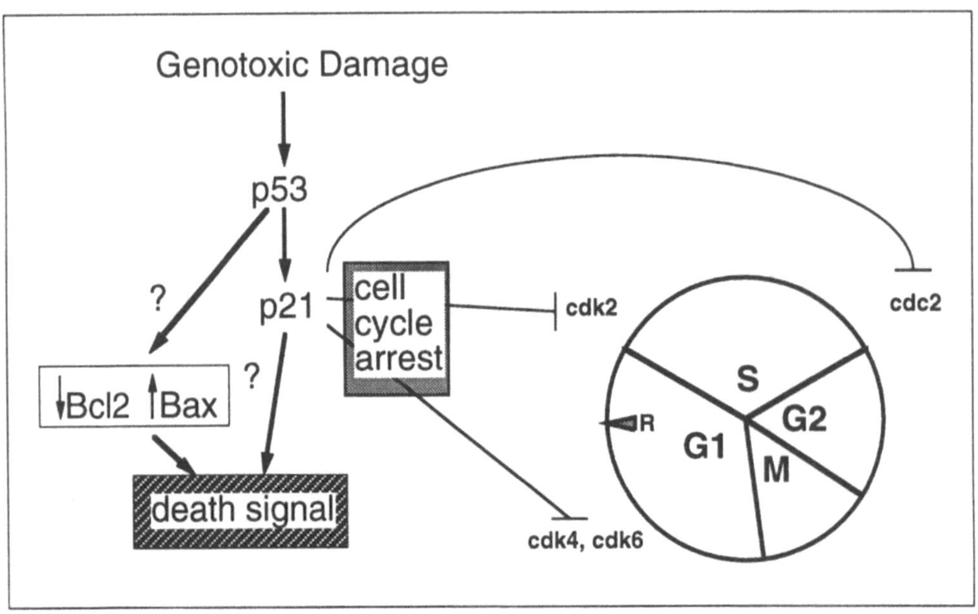

Fig. 1.10. p53 regulates G_1-S phase transition and participates in apoptosis. The tumor suppressor p53 is activated by genotoxic damage such as radiation or certain DNA damaging drugs. p53 in turn activates the transcription of p21 which inhibits many cyclin dependent kinases resulting in cell cycle arrest. p53 also appears to induce the expression of Bax and to inhibit the expression of Bcl-2, thus contributing to the cell death signal. The exact mechanism of p53 induced apoptosis is still being studied.

levels increase (by post-transcriptional events) resulting in G_1 arrest. Overexpression of wild-type p53 protein in tissue culture cells enhances the rate of apoptosis. On the other hand, homologous genetic knockout of the p53 gene resulted in thymocytes that do not undergo apoptosis with exposure to ionizing radiation, but continue to die with exposure to glucocorticoid.[112,113] These observations suggest that there are separate pathways leading to apoptosis in thymocytes, and at least one of these pathways is p53-dependent.

In addition to its role in cell proliferation and neoplastic transformation, a new role of the c-Myc protein in cell fate was discovered when cells overexpressing Myc were deprived of nutrients and growth factors (Fig. 1.11).[118-120] In contrast to parental fibroblastic cell lines that undergo growth arrest and withdraw into the G_0 phase with serum-deprivation, the Myc-overexpressing cells undergo apoptosis. c-Myc and apoptosis is discussed in more depth in chapter 10.

In addition to the nuclear and cytoplasmic effectors of apoptosis, antibody-induced crosslinking of the transmembrane protein termed Fas/APO-1 induces apoptosis. The Fas protein is homologous to receptors for tumor necrosis factor and is involved in cytotoxic T lymphocyte (CTL) mediated apoptosis. CTLs specifically recognize target cells through the T cell receptor complex and lethally engage their targets through the ligand of Fas.[121-124] Mutant mice bearing mutations for either the Fas receptor or its ligand are afflicted with a lymphoproliferative disorder suggesting that cell death is physiological for lymphocytes.[125-127]

1.5 THE TRANSFORMED PHENOTYPE

Normal cell proliferation is orchestrated by signals arising from the cell surface and cytoplasmic messengers that link these afferent signals to gene transcription. Chaos occurring at any step in the control of cell growth or apoptosis can unleash cells from their normal restraints and cause uncontrolled neoplastic cell proliferation. The transformed phenotype is largely due to genetic changes that short-circuit normal pathways involved in normal cell proliferation, although epigenetic changes such as alterations in genomic imprinting also correlate with transformation. Several features of cultured cells have been used as indicators of the transformed phenotype.[128] Among these are: 1) immortalization, 2) decreased

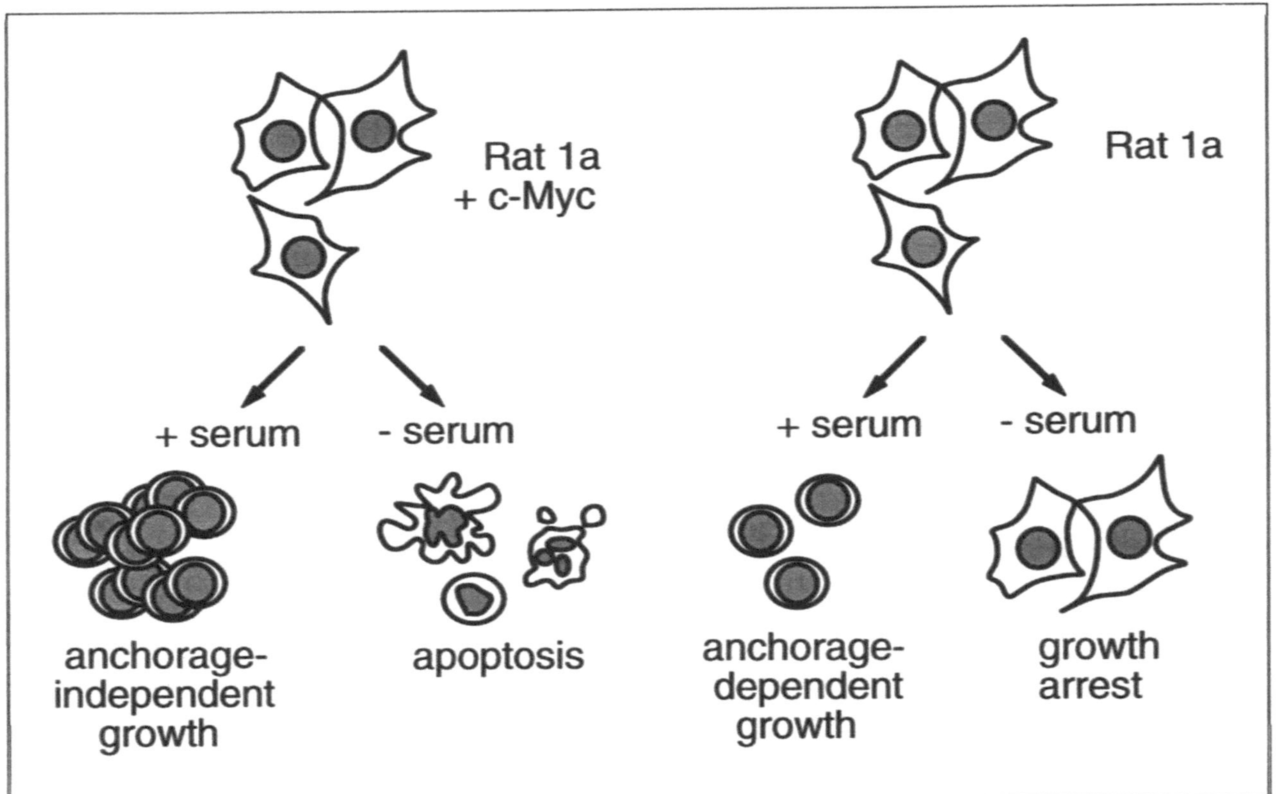

Fig. 1.11. Paradoxical effect of c-Myc on cell proliferation and apoptosis. c-Myc overexpression in specific rat fibroblasts confers anchorage-independent growth in the presence of serum. Serum withdrawal from adherent rat fibroblasts causes them to undergo programmed cell death (left diagram). In contrast, the parental cells are anchorage dependent and serum withdrawal causes growth arrest but not apoptosis (right diagram).

serum dependency, 3) loss of contact growth inhibition and 4) anchorage-independent growth (Fig. 1.12).

Normal primary cells in tissue culture, like the intact organism, has a finite lifespan. Unlike primary cells harvested from animal tissues that senesce in culture (Hayflick phenomenon),[129] transformed cells are immortalized and can be passaged in culture indefinitely. Although the exact molecular mechanisms underlying the cellular clock that tracks the number of cell divisions in normal primary cells is unknown, recent evidence suggests that the shortening of the ends of chromosomes, termed telomeres, is associated with senescence.[130-134] Non-transformed cells in culture require growth factors in serum for optimal growth, whereas transformed cells typically require less serum. Whether

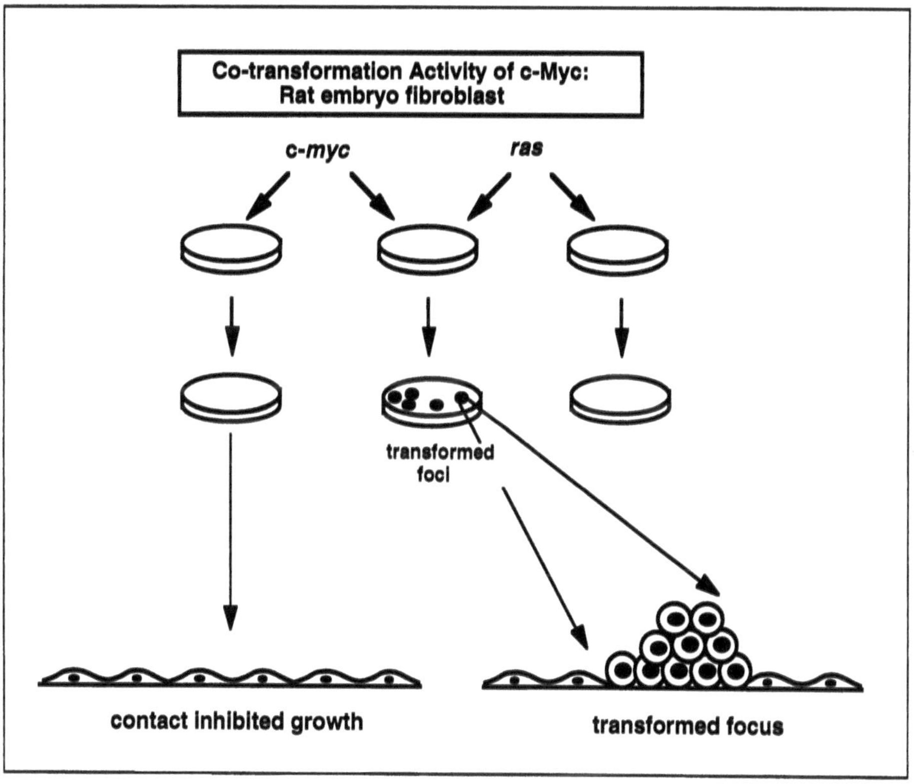

Fig. 1.12. Transformation of primary cells through cooperation of the oncogenes myc and ras. Primary rat embryo cells transfected with either the myc or ras oncogene alone undergo contact inhibition and are untransformed. In contrast, when both myc and ras are expressed, primary embryo cells lose contact inhibition and pile on one another to form transformed foci.

decreased serum requirement is related to autocrine stimulation of growth in transformed cells or short-circuiting of a more downstream signal transduction pathway remains unknown. In the case of normal rat kidney cells transformed by the PDGF-like v-*sis*, there is evidence that v-*sis* transforms via an autocrine pathway.[135]

Normal cells undergo contact inhibition of growth when cells contact each other at high cellular densities. In this situation, cell-cell and cell-extracellular matrix interactions are transduced as growth-inhibitory signals. Transformed cells tend to lose contact inhibition and therefore pile on top of each other at confluency in tissue culture dishes.[136,137] For example, Myc transformation has been studied using the rat embryo cotransformation assay in which cells transfected by c-Myc and activated Ras become transformed and acquire the ability to grow independent of cell contact (Fig. 1.12). Such piles of cells in a culture dish are recognized as foci of transformed cells. Most normal cells proliferate only when they adhere to a substratum, whereas many transformed cells are able to grow in suspension (anchorage-independent growth). Thus, the ability to grow in soft agarose or anchorage-independence is another useful experimental definition of the transformed phenotype.[138] The mechanisms underlying anchorage-dependent growth are beginning to emerge from studies of integrin-mediated adhesion.[139-143] Adhesion through integrin receptors is known to stimulate production of phosphatidylinositol 4,5-bisphosphate (4,5-PIP_2) which has been shown to regulate the function of actin-binding proteins such as gelsolin or profilin. On the other hand, growth factors stimulate the hydrolysis of PIP_2 to generate IP_3, which in turn stimulates protein kinase C and mobilizes intracellular calcium. These signals together constitute normal mitogenic signals stemming from growth factors and contact with the extracellular matrix. Recently, the small GTP-binding protein Rho is implicated in the integrin signalling pathway by its ability to regulate a phosphatidylinositol 4-phosphate 5-kinase that generates PIP_2.[144] Thus anchorage-independent growth of transformed cells may well be the result of deregulation of the Rho mediated pathway in a manner independent of the engagement of the integrin receptor.

Tumorigenicity or the ability of transformed cells to grow in a host animal is one of the best indicators of cellular transformation, although this assay is tedious and expensive. Transformed cells are injected orthotopically (into the organ of origin of the transformed cells) or subcutaneously into syngeneic animals or immunodeficient nude mice and tumor size is determined as an indicator of transformation.

1.6 CONCLUDING REMARKS

With the ease of modern biology techniques, we have rapidly acquired glimpses into the complex processes of cell growth, proliferation and death. In the struggle to understand such multifaceted biological processes, complex phenotypes are undoubtedly reduced to combinations of a linear series of molecular interactions and metamorphoses. Many current experimental approaches have been built on the premise that we can understand the whole by dissecting and understanding the parts of a complicated system. While reductionism is a very successful and often mandatory approach in molecular biology, the enthusiasm for reductionism is tempered by the realization that holism is necessary for full appreciation of biological processes that cannot be recreated simply by reconstituting the isolated components. We have to contend with this limited experimental approach to molecular causality until better experimental approaches are conceived and ground-breaking concepts are framed.

REFERENCES

1. Marshall CJ. Tumor suppressor genes. [Review]. Cell 1991; 64:313-26.
2. Leach FS, Nicolaides NC, Papadopoulos N, et al. Mutations of a mutS homolog in hereditary nonpolyposis colorectal cancer. Cell 1993; 75:1215-25.
3. Papadopoulos N, Nicolaides NC, Wei YF, et al. Mutation of a mutL homolog in hereditary colon cancer [see comments]. Science 1994; 263:1625-9.
4. Bronner CE, Baker SM, Morrison PT, et al. Mutation in the DNA mismatch repair gene homologue hMLH1 is associated with hereditary non-polyposis colon cancer. Nature 1994; 368:258-61.
5. Jiricny J. Colon cancer and DNA repair: have mismatches met their match? [Review]. Trends Genet 1994; 10:164-8.
6. Liu B, Parsons RE, Hamilton SR, et al. hMSH2 mutations in hereditary nonpolyposis colorectal cancer kindreds. Cancer Res 1994; 54:4590-4.

7. Nicolaides NC, Papadopoulos N, Liu B, et al. Mutations of two PMS homologues in hereditary nonpolyposis colon cancer. Nature 1994; 371:75-80.
8. Modrich P. Mismatch repair, genetic stability, and cancer. [Review]. Science 1994; 266:1959-60.
9. Hemminki A, Peltomaki P, Mecklin JP, et al. Loss of the wild type MLH1 gene is a feature of hereditary nonpolyposis colorectal cancer. Nat Genet 1994; 8:405-10.
10. Liu B, Nicolaides NC, Markowitz S, et al. Mismatch repair gene defects in sporadic colorectal cancers with microsatellite instability. Nat Genet 1995; 9:48-55.
11. Bishop JM. Molecular themes in oncogenesis. [Review]. Cell 1991; 64:235-48.
12. Marshall CJ. Specificity of receptor tyrosine kinase signaling: transient versus sustained extracellular signal-regulated kinase activation. [Review]. Cell 1995; 80:179-85.
13. Pazin MJ, Williams LT. Triggering signaling cascades by receptor tyrosine kinases. [Review]. Trends Biochem Sci 1992; 17:374-8.
14. Fantl WJ, Johnson DE, Williams LT. Signalling by receptor tyrosine kinases. [Review]. Annu Rev Biochem 1993; 62:453-81.
15. Divecha N, Irvine RF. Phospholipid signaling. [Review]. Cell 1995; 80:269-78.
16. Clapham DE. Calcium signaling. [Review]. Cell 1995; 80:259-68.
17. Ruff-Jamison S, Chen K, Cohen S. Induction by EGF and interferon-gamma of tyrosine phosphorylated DNA binding proteins in mouse liver nuclei. Science 1993; 261:1733-6.
18. Shibuya H, Yoneyama M, Ninomiya-Tsuji J, Matsumoto K, Taniguchi T. IL-2 and EGF receptors stimulate the hematopoietic cell cycle via different signaling pathways: demonstration of a novel role for c-myc. Cell 1992; 70:57-67.
19. Taniguchi T, Minami Y. The IL-2/IL-2 receptor system: a current overview. [Review]. Cell 1993; 73:5-8.
20. Miyazaki T, Liu ZJ, Kawahara A, et al. Three distinct IL-2 signaling pathways mediated by bcl-2, c-myc, and lck cooperate in hematopoietic cell proliferation. Cell 1995; 81:223-231.
21. Garrell J, Campuzano S. The helix-loop-helix domain: a common motif for bristles, muscles and sex. [Review]. Bioessays 1991; 13:493-8.
22. Tjian R, Maniatis T. Transcriptional activation: a complex puzzle with few easy pieces. [Review]. Cell 1994; 77:5-8.
23. Gill G, Tjian R. Eukaryotic coactivators associated with the TATA box binding protein. [Review]. Curr Opin Genet Dev 1992; 2:236-42.
24. Hill CS, Treisman R. Transcriptional regulation by extracellular signals: mechanisms and specificity. [Review]. Cell 1995; 80:199-211.

25. Serizawa H, Makela TP, Conaway JW, Conaway RC, Weinberg RA, Young RA. Association of Cdk-activating kinase subunits with transcription factor TFIIH. Nature 1995; 374:280-2.
26. Makela TP, Tassan JP, Nigg EA, Frutiger S, Hughes GJ, Weinberg RA. A cyclin associated with the CDK-activating kinase MO15. Nature 1994; 371:254-7.
27. Johnson PF, McKnight SL. Eukaryotic transcriptional regulatory proteins. [Review]. Annu Rev Biochem 1989; 58:799-839.
28. Mitchell PJ, Tjian R. Transcriptional regulation in mammalian cells by sequence-specific DNA binding proteins. [Review]. Science 1989; 245:371-8.
29. Landschulz WH, Johnson PF, McKnight SL. The leucine zipper: a hypothetical structure common to a new class of DNA binding proteins. Science 1988; 240:1759-1764.
30. Murre C, McCaw PS, Baltimore D. A new DNA binding and dimerization motif in immunoglobulin enhancer binding, daughterless, MyoD, and myc proteins. Cell 1989; 56:777-83.
31. Hartwell LH, Kastan MB. Cell cycle control and cancer. [Review]. Science 1994; 266:1821-8.
32. Nurse P. Ordering S phase and M phase in the cell cycle. [Review]. Cell 1994; 79:547-50.
33. Hunter T, Pines J. Cyclins and cancer. [Review]. Cell 1991; 66:1071-4.
34. Hunter T, Pines J. Cyclins and cancer. II: Cyclin D and CDK inhibitors come of age [see comments]. [Review]. Cell 1994; 79:573-82.
35. Geng Y, Weinberg RA. Transforming growth factor beta effects on expression of G1 cyclins and cyclin-dependent protein kinases. Proc Natl Acad Sci USA 1993; 90:10315-9.
36. Sherr CJ. G1 phase progression: cycling on cue [see comments]. [Review]. Cell 1994; 79:551-5.
37. Sherr CJ. Mammalian G1 cyclins. [Review]. Cell 1993; 73:1059-65.
38. Kato JY, Sherr CJ. Inhibition of granulocyte differentiation by G1 cyclins D2 and D3 but not D1. Proc Natl Acad Sci USA 1993; 90:11513-7.
39. Quelle DE, Ashmun RA, Shurtleff SA, et al. Overexpression of mouse D-type cyclins accelerates G1 phase in rodent fibroblasts. Genes Dev 1993; 7:1559-71.
40. Ewen ME, Sluss HK, Sherr CJ, Matsushime H, Kato J, Livingston DM. Functional interactions of the retinoblastoma protein with mammalian D-type cyclins. Cell 1993; 73:487-97.
41. Kato J, Matsushime H, Hiebert SW, Ewen ME, Sherr CJ. Direct binding of cyclin D to the retinoblastoma gene product (pRb) and pRb phosphorylation by the cyclin D-dependent kinase CDK4. Genes Dev 1993; 7:331-42.

42. Flemington EK, Speck SH, Kaelin W, Jr. E2F-1-mediated transactivation is inhibited by complex formation with the retinoblastoma susceptibility gene product. Proc Natl Acad Sci USA 1993; 90:6914-8.
43. Krek W, Livingston DM, Shirodkar S. Binding to DNA and the retinoblastoma gene product promoted by complex formation of different E2F family members. Science 1993; 262:1557-1560.
44. Shirodkar S, Ewen M, DeCaprio JA, Morgan J, Livingston DM, Chittenden T. The transcription factor E2F interacts with the retinoblastoma product and a p107-cyclin A complex in a cell cycle-regulated manner. Cell 1992; 68:157-66.
45. Mudryj M, Devoto SH, Hiebert SW, Hunter T, Pines J, Nevins JR. Cell cycle regulation of the E2F transcription factor involves an interaction with cyclin A. Cell 1991; 65:1243-53.
46. Lees E, Faha B, Dulic V, Reed S, Harlow E. Cylin E/cdk2 and cyclin A/cdk2 kinases associate with p107 and E2F in a temporally distinct manner. Genes Dev 1992; 6:1874-1885.
47. Resnitzky D, Gossen M, Bujard H, Reed SI. Acceleration of the G1/S phase transition by expression of cyclins D1 and E with an inducible system. Mol Cell Biol 1994; 14:1669-79.
48. Koff A, Giordano A, Desai D, et al. Formation and activation of a cyclin E-cdk2 complex during the G1 phase of the human cell cycle. Science 1992; 257:1689-94.
49. Koff A, Ohtsuki M, Polyak K, Roberts JM, Massague J. Negative regulation of G1 in mammalian cells: inhibition of cyclin E-dependent kinase by TGF-beta. Science 1993; 260:536-9.
50. Ohtsubo M, Roberts JM. Cyclin-dependent regulation of G1 in mammalian fibroblasts. Science 1993; 259:1908-12.
51. Kastan MB, Zhan Q, el-Deiry WS, et al. A mammalian cell cycle checkpoint pathway utilizing p53 and GADD45 is defective in ataxia-telangiectasia. Cell 1992; 71:587-97.
52. Hunter T. Braking the cycle. [Review]. Cell 1993; 75:839-41.
53. Polyak K, Kato JY, Solomon MJ, et al. p27Kip1, a cyclin-Cdk inhibitor, links transforming growth factor-beta and contact inhibition to cell cycle arrest. Genes Dev 1994; 8:9-22.
54. Toyoshima H, Hunter T. p27, a novel inhibitor of G1 cyclin-Cdk protein kinase activity, is related to p21. Cell 1994; 78:67-74.
55. Kamb A, Gruis NA, Weaver-Feldhaus J, et al. A cell cycle regulator potentially involved in genesis of many tumor types. Science 1994; 264:436-40.
56. Cairns P, Mao L, Merlo A, et al. Rates of p16 (MTS1) mutations in primary tumors with 9p loss [letter; comment]. Science 1994; 265:415-7.
57. Marx J. A challenge to p16 gene as a major tumor suppressor [news]. Science 1994; 264.
58. Spruck Cr, Gonzalez-Zulueta M, Shibata A, et al. p16 gene in uncultured tumours [letter] [see comments]. Nature 1994; 370:183-4.

59. Pines J, Hunter T. Human cyclin A is adenovirus E1A-associated protein p60 and behaves differently from cyclin B. Nature 1990; 346:760-3.
60. Faha B, Ewen ME, Tsai LH, Livingston DM, Harlow E. Interaction between human cyclin A and adenovirus E1A-associated p107 protein. Science 1992; 255:87-90.
61. Girard F, Strausfeld U, Fernandez A, Lamb NJ. Cyclin A is required for the onset of DNA replication in mammalian fibroblasts. Cell 1991; 67:1169-79.
62. Pagano M, Pepperkok R, Verde F, Ansorge W, Draetta G. Cyclin A is required at two points in the human cell cycle. EMBO J 1992; 11:961-71.
63. Sobczak-Thepot J, Harper F, Florentin Y, Zindy F, Brechot C, Puvion E. Localization of cyclin A at the sites of cellular DNA replication. Exp Cell Res 1993; 206:43-8.
64. Devoto SH, Mudryj M, Pines J, Hunter T, Nevins JR. A cyclin A-protein kinase complex possesses sequence-specific DNA binding activity: p33cdk2 is a component of the E2F-cyclin A complex. Cell 1992; 68:167-76.
65. Peeper DS, Parker LL, Ewen ME, et al. A- and B-type cyclins differentially modulate substrate specificity of cyclin-cdk complexes. EMBO J 1993; 12:1947-54.
66. Marraccino RL, Firpo EJ, Roberts JM. Activation of the p34 CDC2 protein kinase at the start of S phase in the human cell cycle. Mol Biol Cell 1992; 3:389-401.
67. Gorbsky GJ. Chromosome Motion in Mitosis. BioEssays 1992; 14:73-80.
68. King RW, Jackson PK, Kirschner MW. Mitosis in transition. [Review]. Cell 1994; 79:563-71.
69. Kuriyama R, Nislow C. Molecular Components of the Mitotic Spindle. BioEssays 1992; 14:81-88.
70. Vasseur-Cognet M, Lane MD. Trans-acting factors involved in adipogenic differentiation. [Review]. Curr Opin Genet Dev 1993; 3:238-45.
71. Freytag SO, Paielli DL, Gilbert JD. Ectopic expression of the CCAAT/enhancer-binding protein alpha promotes the adipogenic program in a variety of mouse fibroblastic cells. Genes Dev 1994; 8:1654-63.
72. Freytag SO, Geddes TJ. Reciprocal regulation of adipogenesis by Myc and C/EBP alpha. Science 1992; 256:379-82.
73. Olson EN. Proto-oncogenes in the regulatory circuit for myogenesis. [Review]. Semin Cell Biol 1992; 3:127-36.
74. Weintraub H. The MyoD family and myogenesis: redundancy, networks, and thresholds. [Review]. Cell 1993; 75:1241-4.
75. Atchley WR, Fitch WM, Bronner-Fraser M. Molecular evolution of the MyoD family of transcription factors. Proc Natl Acad Sci USA 1994; 91:11522-6.

76. Weintraub H, Davis R, Tapscott S, et al. The myoD gene family: nodal point during specification of the muscle cell lineage. [Review]. Science 1991; 251:761-6.
77. Lassar AB, Paterson BM, Weintraub H. Transfection of a DNA locus that mediates the conversion of 10T1/2 fibroblasts to myoblasts. Cell 1986; 47:649-56.
78. Davis RL, Weintraub H, Lassar AB. Expression of a single transfected cDNA converts fibroblasts to myoblasts. Cell 1987; 51:987-1000.
79. Hasty P, Bradley A, Morris JH, et al. Muscle deficiency and neonatal death in mice with a targeted mutation in the myogenin gene [see comments]. Nature 1993; 364:501-6.
80. Rudnicki MA, Braun T, Hinuma S, Jaenisch R. Inactivation of MyoD in mice leads to up-regulation of the myogenic HLH gene Myf-5 and results in apparently normal muscle development. Cell 1992; 71:383-90.
81. Murre C, McCaw PS, Vaessin H, et al. Interactions between heterologous helix-loop-helix proteins generate complexes that bind specifically to a common DNA sequence. Cell 1989; 58:537-44.
82. Benezra R, Davis RL, Lockshon D, Turner DL, Weintraub H. The protein Id: a negative regulator of helix-loop-helix DNA binding proteins. Cell 1990; 61:49-59.
83. Finkel T, Duc J, Fearon ER, Dang CV, Tomaselli GF. Detection and modulation in vivo of helix-loop-helix protein-protein interactions. J Biol Chem 1993; 268:5-8.
84. Orkin SH. Transcription factors and hematopoietic development. [Review]. J Biol Chem 1995; 270:4955-8.
85. Weiss MJ, Orkin SH. GATA transcription factors: key regulators of hematopoiesis. [Review]. Exp Hematol 1995; 23:99-107.
86. Pevny L, Lin CS, D'Agati V, Simon MC, Orkin SH, Costantini F. Development of hematopoietic cells lacking transcription factor GATA-1. Development 1995; 121:163-72.
87. Tsai FY, Keller G, Kuo FC, et al. An early haematopoietic defect in mice lacking the transcription factor GATA-2. Nature 1994; 371:221-6.
88. Merika M, Orkin SH. DNA-binding specificity of GATA family transcription factors. Mol Cell Biol 1993; 13:3999-4010.
89. Orkin SH. GATA-binding transcription factors in hematopoietic cells. [Review]. Blood 1992; 80:575-81.
90. Gavrieli Y, Sherman Y, Ben-Sasson SA. Identification of programmed cell death in situ via specific labeling of nuclear DNA fragmentation. J Cell Biol 1992; 119:493-501.
91. Ellis RE, Yuan JY, Horvitz HR. Mechanisms and functions of cell death. [Review]. Annu Rev Cell Biol 1991; 7:663-98.
92. Raff MC. Social controls on cell survival and cell death. [Review]. Nature 1992; 356:397-400.

93. Vaux DL. Toward an understanding of the molecular mechanisms of physiological cell death. [Review]. Proc Natl Acad Sci USA 1993; 90:786-9.
94. Kerr JF, Wyllie AH, Currie AR. Apoptosis: a basic biological phenomenon with wide-ranging implications in tissue kinetics. [Review]. Br J Cancer 1972; 26:239-57.
95. Wyllie AH. Death in normal and neoplastic cells. [Review]. J Clin Pathol Suppl 1974; 7:35-42.
96. Hengartner MO, Horvitz HR. Programmed cell death in Caenorhabditis elegans. [Review]. Curr Opin Genet Dev 1994; 4:581-6.
97. Yuan J, Shaham S, Ledoux S, Ellis HM, Horvitz HR. The C. elegans cell death gene ced-3 encodes a protein similar to mammalian interleukin-1 beta-converting enzyme. Cell 1993; 75:641-52.
98. Yuan J, Horvitz HR. The Caenorhabditis elegans cell death gene ced-4 encodes a novel protein and is expressed during the period of extensive programmed cell death. Development 1992; 116:309-20.
99. Jacobson MD, Evan GI. Apoptosis. Breaking the ICE. [Review]. Curr Biol 1994; 4:337-40.
100. Lazebnik YA, Kaufmann SH, Desnoyers S, Poirier GG, Earnshaw WC. Cleavage of poly(ADP-ribose) polymerase by a proteinase with properties like ICE. Nature 1994; 371:346-7.
101. Irmler M, Hertig S, MacDonald HR, et al. Granzyme A is an interleukin 1 beta-converting enzyme. J Exp Med 1995; 181:1917-22.
102. Hengartner MO, Ellis RE, Horvitz HR. Caenorhabditis elegans gene ced-9 protects cells from programmed cell death. Nature 1992; 356:494-9.
103. Hengartner MO, Horvitz HR. C. elegans cell survival gene ced-9 encodes a functional homolog of the mammalian proto-oncogene bcl-2. Cell 1994; 76:665-76.
104. Hengartner MO, Horvitz HR. Activation of C. elegans cell death protein CED-9 by an amino-acid substitution in a domain conserved in Bcl-2. Nature 1994; 369:318-20.
105. Korsmeyer SJ. Bcl-2: an antidote to programmed cell death. [Review]. Cancer Surv 1992; 15:105-18.
106. Hockenbery DM, Oltvai ZN, Yin XM, Milliman CL, Korsmeyer SJ. Bcl-2 functions in an antioxidant pathway to prevent apoptosis. Cell 1993; 75:241-51.
107. Korsmeyer SJ, Shutter JR, Veis DJ, Merry DE, Oltvai ZN. Bcl-2/Bax: a rheostat that regulates an anti-oxidant pathway and cell death. [Review]. Semin Cancer Biol 1993; 4:327-32.
108. Jacobson MD, Raff MC. Programmed cell death and Bcl-2 protection in very low oxygen. Nature 1995; 374:814-6.

109. Yang E, Zha J, Jockel J, Boise LH, Thompson CB, Korsmeyer SJ. Bad, a heterodimeric partner for Bcl-XL and Bcl-2, displaces Bax and promotes cell death. Cell 1995; 80:285-91.
110. Yin XM, Oltval ZN, Korsmeyer SJ. BH1 and BH2 domains of Bcl-2 are required for inhibition of apoptosis and heterodimerization with Bax [see comments]. Nature 1994; 369:321-3.
111. Oltvai ZN, Milliman CL, Korsmeyer SJ. Bcl-2 heterodimerizes in vivo with a conserved homolog, Bax, that accelerates programmed cell death. Cell 1993; 74:609-19.
112. Clarke AR, Purdie CA, Harrison DJ, et al. Thymocyte apoptosis induced by p53-dependent and independent pathways. Nature 1993; 362:849-52.
113. Lowe SW, Schmitt EM, Smith SW, Osborne BA, Jacks T. p53 is required for radiation-induced apoptosis in mouse thymocytes. Nature 1993; 362:847-9.
114. Canman CE, Gilmer TM, Coutts SB, Kastan MB. Growth factor modulation of p53-mediated growth arrest versus apoptosis. Genes Dev 1995; 9:600-11.
115. Morgenbesser SD, Williams BO, Jacks T, DePinho RA. p53-dependent apoptosis produced by Rb-deficiency in the developing mouse lens. Nature 1994; 371:72-4.
116. Ryan JJ, Prochownik E, Gottlieb CA, et al. C-myc and bcl-2 modulate p53 function by altering p53 subcellular trafficking during the cell cycle. Proc Natl Acad Sci USA 1994; 91:5878-82.
117. Hermeking H, Eick D. Mediation of c-myc-induced apoptosis by p53. Science 1994; 265:2091-3.
118. Evan GI, Wyllie AH, Gilbert CS, et al. Induction of apoptosis in fibroblasts by c-myc protein. Cell 1992; 69:119-28.
119. Evan G, Harrington E, Fanidi A, Land H, Amati B, Bennett M. Integrated control of cell proliferation and cell death by the c-myc oncogene. Proc R Soc Lond B Biol Sci 1994; 345:269-75.
120. Askew DS, Ashmun RA, Simmons BC, Cleveland JL. Constitutive c-myc expression in an IL-3-dependent myeloid cell line suppresses cell cycle arrest and accelerates apoptosis. Oncogene 1991; 6:1915-22.
121. Itoh N, Yonehara S, Ishii A, et al. The polypeptide encoded by the cDNA for human cell surface antigen Fas can mediate apoptosis. Cell 1991; 66:233-43.
122. Suda T, Takahashi T, Golstein P, Nagata S. Molecular cloning and expression of the Fas ligand, a novel member of the tumor necrosis factor family. Cell 1993; 75:1169-78.
123. Nagata S, Golstein P. The Fas death factor. [Review]. Science 1995; 267:1449-56.
124. Alderson MR, Tough TW, Davis-Smith T, et al. Fas ligand mediates activation-induced cell death in human T lymphocytes. J Exp Med 1995; 181:71-7.

125. Takahashi T, Tanaka M, Brannan CI, et al. Generalized lymphoproliferative disease in mice, caused by a point mutation in the Fas ligand. Cell 1994; 76:969-76.
126. Singer GG, Carrera AC, Marshak-Rothstein A, Martinez C, Abbas AK. Apoptosis, Fas and systemic autoimmunity: the MRL-lpr/lpr model. [Review]. Curr Opin Immunol 1994; 6:913-20.
127. Nagata S, Suda T. Fas and Fas ligand: lpr and gld mutations. [Review]. Immunol Today 1995; 16:39-43.
128. Cox AD, Der CJ. Biological assays for cellular transformation. Methods Enzymol 1994; 238:277-94.
129. Hayflick L. The limited in vitro lifetime of human diploid cell strains. Exp Cell Res 1965; 37:614-36.
130. Harley CB, Vaziri H, Counter CM, Allsopp RC. The telomere hypothesis of cellular aging. Exp Gerontol 1992; 27:375-82.
131. Allsopp RC, Vaziri H, Patterson C, et al. Telomere length predicts replicative capacity of human fibroblasts. Proc Natl Acad Sci USA 1992; 89:10114-8.
132. Levy MZ, Allsopp RC, Futcher AB, Greider CW, Harley CB. Telomere end-replication problem and cell aging. J Mol Biol 1992; 225:951-60.
133. Kennedy BK, Austriaco N, Jr., Zhang J, Guarente L. Mutation in the silencing gene SIR4 can delay aging in S. cerevisiae. Cell 1995; 80:485-96.
134. Vaziri H, Schachter F, Uchida I, et al. Loss of telomeric DNA during aging of normal and trisomy 21 human lymphocytes. Am J Hum Genet 1993; 52:661-7.
135. Keating MT, Williams LT. Autocrine stimulation of intracellular PDGF receptors in v-sis-transformed cells. Science 1988; 239:914-6.
136. Temin HM, Rubin H. Characteristic of an assay for Rous Sarcoma Virus and Rous sarcoma cells in tissue culture. Virology 1958; 6:669-88.
137. Todaro GJ, Green H. An assay for cellular transformation by SV40. Virology 1964; 23:117-9.
138. Macpherson I, Montaignier L. Agar suspension culture for the selective assay of cells transformed by polyoma virus. Virology 1964; 23:291-4.
139. Schwartz MA, Ingber DE. Integrating with integrins. [Review]. Mol Biol Cell 1994; 5:389-93.
140. Schwartz MA, Denninghoff K. Alpha v integrins mediate the rise in intracellular calcium in endothelial cells on fibronectin even though they play a minor role in adhesion. J Biol Chem 1994; 269:11133-7.
141. Meredith J, Jr., Fazeli B, Schwartz MA. The extracellular matrix as a cell survival factor. Mol Biol Cell 1993; 4:953-61.

142. McNamee HP, Ingber DE, Schwartz MA. Adhesion to fibronectin stimulates inositol lipid synthesis and enhances PDGF-induced inositol lipid breakdown. J Cell Biol 1993; 121:673-8.
143. Schwartz MA. Signaling by integrins: implications for tumorigenesis. Cancer Res 1993; 53:1503-6.
144. Chong LD, Traynor-Kaplan A, Bokoch GM, Schwartz MA. The small GTP-binding protein Rho regulates a phosphatidylinositol 4-phosphate 5-kinase in mammalian cells. Cell 1994; 79:507-13.

CHAPTER 2

Retroviruses, Cancer Genes, and Tumor Suppressor Genes

2.1 ORIGINS OF ONCOGENES

Cellular transformation may occur spontaneously as a result of errors in cellular safeguard mechanisms such as faulty DNA repair, exposure to physical and chemical mutagens or infection by tumorigenic viruses. Studies of tumorigenic DNA and RNA viruses have contributed profoundly to our understanding of molecular oncogenesis, since these viruses contribute to tumor formation by usurping the machineries involved in the control of cell proliferation. Some of the RNA retroviruses transduce captured cellular proto-oncogenes termed viral oncogenes which have undergone mutations. Other retroviruses can transform in the absence of a viral oncogene by inserting strong transcriptional activating elements near cellular genes that contribute positively to cell growth. The DNA tumor viruses often carry viral genes that are oncogenic.

2.1.1 DNA Tumor Viruses

Adenovirus, papovavirus and herpesvirus are among some of the better studied DNA tumor viruses. Many of these viruses encode key viral proteins that interfere with cellular growth regulatory proteins. The simian virus 40 (SV40) genome encodes a protein called large tumor (T) antigen that contributes to the tumorigenic properties of SV40 (Fig. 2.1).[1-5] The T antigen is a 708 amino acid phosphoprotein that participates in SV40 DNA replication and gene transcription. It is capable of binding to and inactivating the tumor suppressor proteins p53 and pRb, which play important roles in the progression of the cell cycle (see chapter 1).[1,5-8] Likewise, the adenoviral E1A protein binds pRb as well

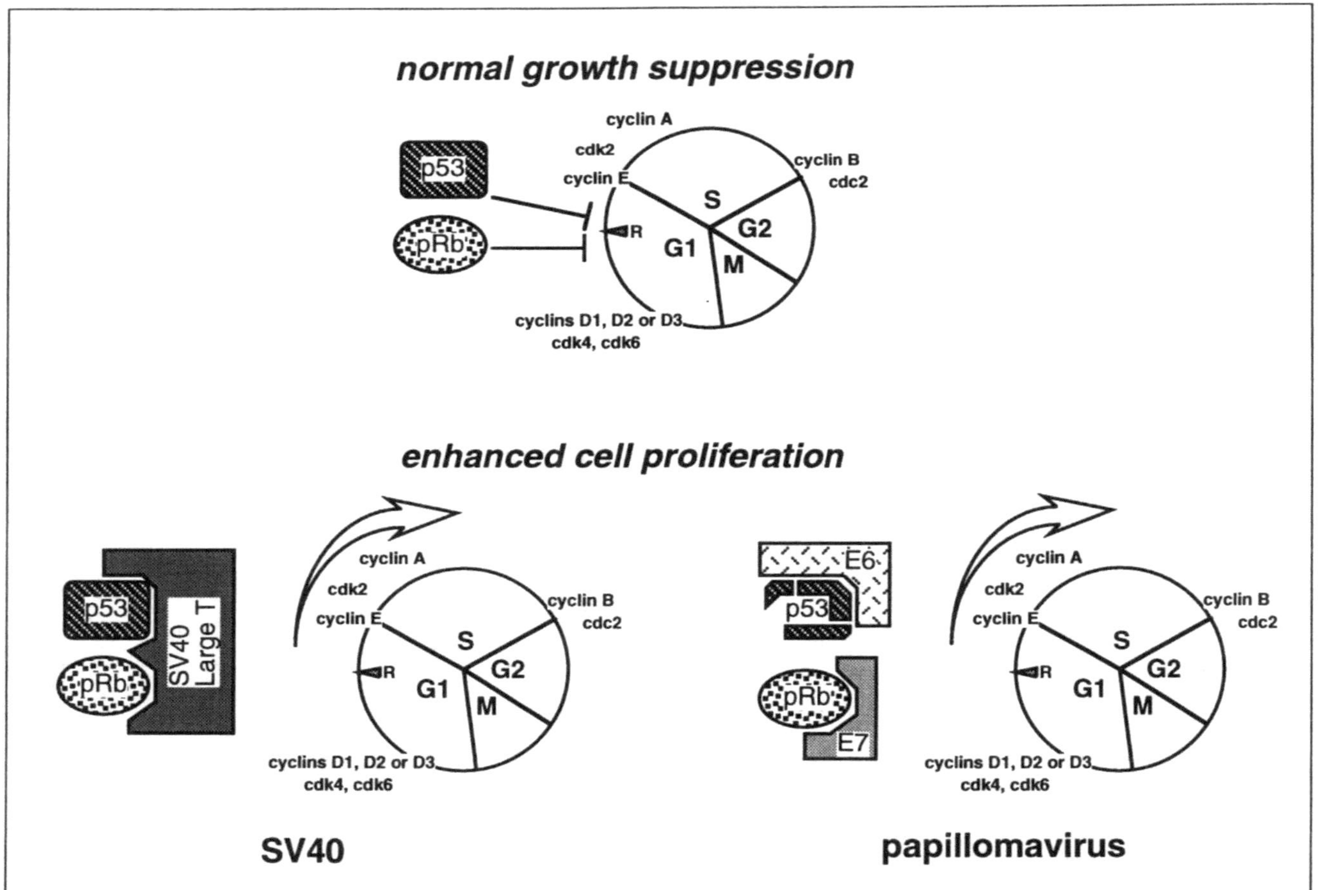

Fig. 2.1. Oncogenic DNA viruses inactivate p53 and pRb via viral proteins. p53 and pRb negatively regulate the transition of normal cells from G_1 into S phase of the cell cycle (top diagram). The SV40 large T antigen physically binds both p53 and pRb so that they are incapable of inhibiting cell proliferation. The papillomaviral E6 protein, which is similar to a component of the ubiquitin protein degradation pathway, binds p53 and primes it for proteolytic degradation. The papillomaviral E7 protein also physically binds pRb and blocks its activity.

as the pRb-related protein p107.1. E1A also affects transcriptional activation by interacting with the octamer binding transcription factors and by interacting with the basal transcription factor, TFIIB. The adenoviral E1B p55 protein, which cooperates with E1A to transform cells, is able to bind to and inactivate p53,[9,10] whereas the E1B p19 protein protect cells against programmed cell death.[11-15]

Oncogenic human papillomaviruses, which contribute to the development of human cervical cancer, encode a protein termed E6 that enhances the degradation of the tumor suppressor p53 (Fig. 2.1).[16-18] It also encodes the E7 protein that binds to and inactivate pRb and p107.[19-21] In contrast, the bovine papillomavirus has been shown to directly activate the platelet-derived growth factor receptor through a direct interaction with one of its proteins termed E5. The hepatitis B hepadnavirus contributes to the genesis of hepatoma in part through the hepatitis B pX protein that behaves as a transcriptional co-activator by binding to cellular transcriptional regulatory proteins.[22,23] Although the details of the molecular mechanisms of oncogenesis by DNA tumor viruses are only beginning to emerge, it has become apparent that many of these viruses have acquired proteins that undermine cellular growth regulators, such as p53 and pRb, to induce tumor formation.

2.1.2 RNA Tumor Viruses

The origin of investigations on RNA (retrovirus) tumor viruses began with Peyton Rous's 1910 discovery. He found that avian sarcomas could be transmitted horizontally by submicroscopic filterable (viruses) agents isolated from a chicken sarcoma. By 1970, isolated Rous sarcoma virus was shown to induce focus formation (i.e. transformation) of rodent or chicken cells in culture. Many oncogenic retroviruses have since been isolated from a variety of naturally occurring animal tumors. Curiously, such retroviruses bearing transduced oncogenes have not been identified in human cancers. Nevertheless, the viral oncogenes have been frequently connected with some human neoplastic condition.[24-27]

In general, typical retroviruses have three protein-coding genes: *gag*, *pol*, and *env* (Fig. 2.2). The retrovirion is comprised of an envelope derived from the plasma membrane of the host cell that contains the product of the *env* gene and encloses a nucleoprotein core containing capsid proteins encoded by the *gag* gene.[28] The genetic organization of retroviral genomes has been studied in great detail. The RNA tumor viruses are classified as acutely or chronically transformed types.

The acutely transforming viruses carry an incomplete genome that is partially replaced by a viral oncogene. Except for the Rous sarcoma virus, oncogenic retroviruses require helper viruses to complement their defective genome for viral replication. Viral oncogenes originate from transduced host cellular genes or proto-oncogenes that are captured by the viral genome presumably during transcription and recombination events (Fig. 2.2). The first evidence for this paradigm came from the work of Bishop and

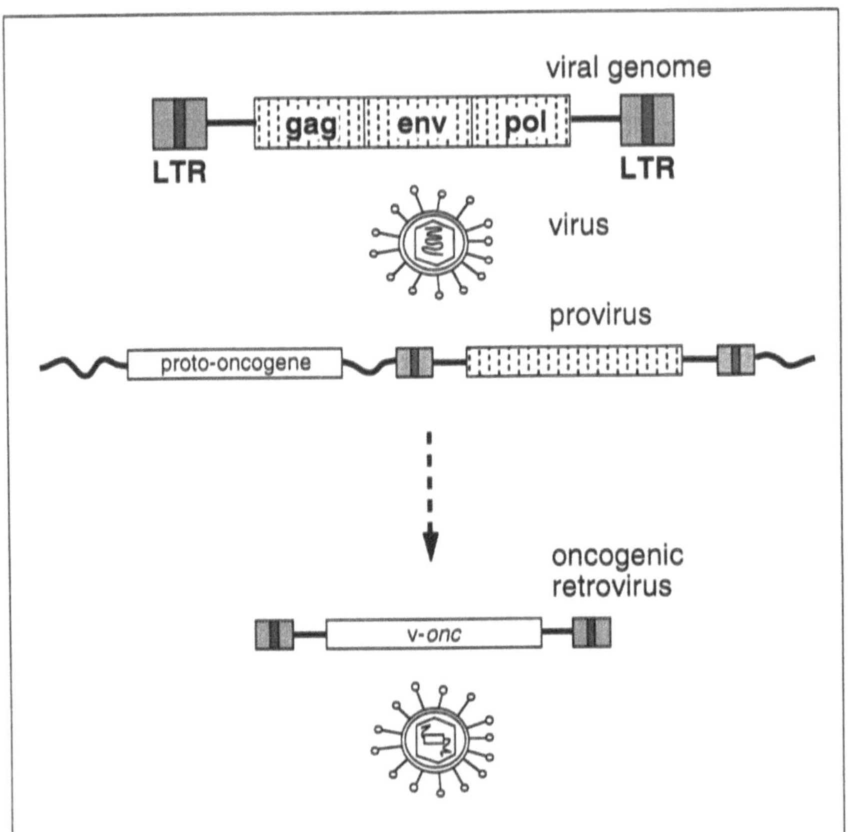

Fig. 2.2. Retroviral transduction of oncogenes. A generic retroviral genome is depicted at the top with viral genes gag, env, and pol. The membrane encapsulated virus infects cells and integrates into the host genome as a provirus (middle diagram). Errors in transcription of the proviral genome that capture surrounding host DNA culminate in defective viruses bearing transduced host DNA. Host DNA capable of inducing neoplasia are propagated (by helper viruses in certain cases) and recognized as viral oncogenes, v-onc.

Varmus with the Rous sarcoma virus.[27] While the homologous proto-oncogenes are non-oncogenic, the retroviral oncogenes are potently transforming and are notably altered from their cellular proto-oncogenic counterpart. For example, many of the viral oncoproteins are fused to the viral *gag* product. Coding sequence alterations are frequently found in the oncogene encoded portions in comparison to the corresponding proto-oncogene. In a variety of retroviruses, the viral oncogene is sufficient for transformation suggesting that genetic alterations in the oncogenes have been selected for their potency to transform cells, a property that allows them to propagate.

In contrast, chronically transforming retroviruses, whose incubation periods usually span months, contain intact viral genomes. They transform cells by insertional activation of proto-oncogenes. Invariably, insertional activation of cellular genes has been attributed to strong transcriptional activating elements in the retroviral regulatory sequences termed the long terminal repeat (LTR). For example, the Moloney murine leukemia virus causes T-cell lymphoma in the mouse by insertional activation of the c-*myc* gene.[29,30]

2.2 ONCOGENES: ABERRATIONS OF SIGNAL TRANSDUCTION

Activation of proto-oncogenes to potentially lethal oncogenes is like the awakening of the enemy from within.[31-34] The mechanisms by which normal cellular proto-oncogenes are unleashed to turn against the organism are now emerging. Much about oncogene activation has been learned from studies of viral oncogenes. Activation processes can be viewed as molecular defects that short-circuit the normal pathways controlling cell proliferation. A variety of modes of activation have been recognized depending on the function of the protein encoded by the oncogene. Oncogenic defects at almost all levels of control of cell proliferation have been identified as will be described in the following section.

2.2.1 Growth Factors

Aberrant, neoplastic cell growth can result from the deregulated expression of genes encoding growth factors or structurally altered growth-factor-like substances. Such deregulated expression in cells that bear receptors for the growth factor could trap cells in a relentless positive feedback loop known as autocrine stimulation. The v-*sis* oncogene causes sarcomas in cats and monkeys and encodes a p28 product that was found to have striking homology with the β-chain of the platelet-derived growth factor (PDGF).[35] Fibroblasts transformed by v-*sis*, indeed, exhibit an intracellular autocrine loop that contributes to their transformed phenotype.[36-39] Oncogenic alleles of PDGF have not been identified in human beings, however.

The oncogene *wnt-1* is insertionally activated by the MMTV virus in murine mammary carcinomas.[40,41] The *wnt-1* protein exhibits striking homology to the predicted amino acid sequence of the *Drosophila wingless* gene product that is involved in the normal development of fruit fly appendages.[42,43] Wnt-1 protein is a glycoprotein that behaves like a secreted paracrine growth factor.[44-51] Another MMTV insertionally activated gene *wnt-2* encodes a protein that is also homologous to FGF,[55,57] the transforming gene isolated from Kaposi sarcoma,[52,53] and that isolated from human stomach cancer, *hst*.[54-56] Such striking homology of certain oncoproteins to growth factors suggests that when aberrantly expressed or mutated, growth factors can profoundly affect tumorigenesis.

2.2.2 Growth Factor Receptors

Normal growth factor receptors transduce signals originating from an embrace of growth factors by their extracellular receptor domains. Transduction through the transmembrane domain of the receptor is accomplished by allosteric changes and often by receptor dimerization. Several retroviral oncogenes best exemplify how normal receptors can be converted to oncogenic membrane proteins. The v-*erbB* oncogene found in avian erythroblastosis encodes a truncated version of the epidermal growth factor receptor that lacks the extracellular ligand binding domain.[58] The v-ErbB protein is constitutively activated presumably due to the lack of a domain that negatively regulates its cytoplasmic tyrosine kinase domain. The c-*neu* (*erb*B2) oncogene encodes an EGF-receptor related product called Neu that is activated by a point mutation in the transmembrane domain caused by exposure of rats to the mutagen ethylnitrosourea.[59-61] The resultant oncogene contributes to the development of rat neuroblastoma. An *erb*B2 homolog has been found to be amplified in about 25% of human breast cancers.[62-65] The ligand for Neu was recently identified and termed Heregulin.[66-68]

Genes that encode key growth factors or growth factor receptors in hematopoiesis have been usurped by retroviruses as oncogenes. The feline sarcoma virus (FeSV) transduces the oncogene v-*fms* whose product resembles the colony stimulating factor-1 (CSF-1 or M-CSF) receptor.[69-73] CSF-1 stimulates the growth of hematopoietic cells committed to the monocyte-macrophage lineage. The v-Fms protein has a C-terminal truncation which removes a tyrosine phosphorylation site that regulates the c-Fms tyrosine kinase activity. This alteration causes v-Fms to be constitutively activated. While c-Fms appears to be restricted to the monocyte-macrophage lineage, transgenic mice expressing v-Fms display multilineage abnormalities.

Occasionally in science different fields of investigations converge in a tunnel of inquiry and emerge with profoundly novel connections. The story of *kit* is such an example in which a well-studied retrovirus is linked to two heritable anemias in two well-studied mouse strains bearing either the *steel* or w^v/w^v defect.[74-78] The Hardy-Zuckerman 4 feline sarcoma virus bears the oncogene v-*kit* that encodes the receptor for stem cell factor (SCF). The *steel* and w^v/w^v mutant mice lack functional SCF and SCF receptor, respectively. SCF figures prominently among early lineage nonspecific hematopoietic growth factors that participate in early hematopoiesis.

2.2.3 Nonreceptor Protein Tyrosine Kinase

Another group of oncoproteins that serve to link the extracellular world with the nucleus is the *src* family of tyrosine kinases. Some of these proteins associate with the plasma membrane through hydrophobic fatty acid tails tacked onto the proteins by myristylation, for example. The Src protein itself contains the prototypic Src-homology domain 2 (SH2 domain) that is a short sequence of 100 amino acids capable of binding to phosphotyrosine peptide sequences.[79,80] Through the SH2 domain, Src is presumably capable of binding to cytoplasmic tails of the receptor tyrosine kinases which autophosphorylate upon activation. The C-terminal domain of Src inhibits the SH2 domain unless the C-terminus is phosphorylated. v-Src, the protein encoded by the retrovirally transduced oncogene, lacks this inhibitory domain. Thus, v-Src may be constitutively activated as a result of this acquired mutation. Protein substrates that are phosphorylated by Src have been identified but their connection to transformation by Src has been less well established.

Bcr-Abl is another well-known protein tyrosine kinase that is strongly linked to the pathogenesis of human chronic myelogenous leukemia (CML).[81-95] c-*abl*, a proto-oncogene which encodes a tyrosine kinase, is activated by t(9:22) chromosomal translocation found in ninety percent of CML cases. This translocation fuses the amino-terminal coding sequence of the breakpoint cluster region (*bcr*) to exon 2 of c-*abl*. The resultant fusion protein contains both SH2 and SH3 domains of c-Abl but with augmented tyrosine kinase activity which accounts for its transformation ability.[96,97] While the normal counterpart, c-Abl, localizes to the nucleus, Bcr-Abl localizes to the cytoplasm probably as a result of losing the nuclear targeting sequence. One pathway by which Bcr-Abl signals the cell nucleus is through the Ras-Jun kinase signal transduction pathway which ultimately may affect the activity of transcription factors like Jun or c-Myc.

2.2.4 Guanine Nucleotide Exchange Proteins

The Ras proteins represent a unique family of oncoproteins that appear sufficient for cellular transformation when activated by mutations (Fig. 2.3).[98-105] Several of the mutated *ras* genes are frequently found associated with human cancers.[101,106,107] These proteins associate with membrane receptors in the cytoplasmic com-

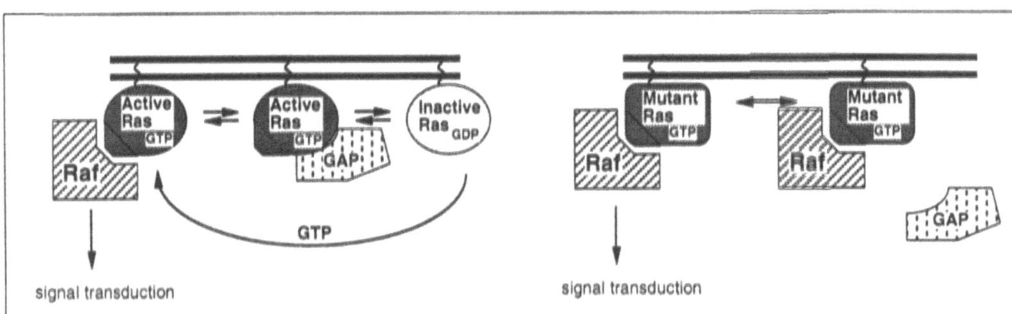

Fig. 2.3. The Ras pathway of signal transduction. The GTP binding proteins, of which Ras is a member, transduce signals from receptors that span the plasma membrane seven times (not shown). Active Ras bound to GTP interacts with Raf kinase to transduce signals through the MAP kinase pathway. The intrinsic GTPase activity of Ras is augmented by its interaction with the GTPase activation protein (GAP) resulting in the hydrolysis of GTP to GDP and inactivation of Ras. Mutated oncogenic Ras remains bound to GTP, resists interaction with GAP and results in constitutive activation of Raf.

partment and participate in transducing extracellular signals to the cell nucleus. The connection between certain growth factor receptors and the Ras signalling pathway is achieved through interactions between intermediary proteins such as Sos and Grb2 (Fig. 1.1). For example, when the epidermal growth factor tyrosine kinase receptor (EGFR) engages its ligand EGF, autophosphorylation of the receptor occurs. The phosphotyrosine on the EGFR recruits Grb2, which is an SH2 containing protein that in turn, binds Son of sevenless (Sos).[108-112] Sos directly binds Ras and passes the signal on from the extracellular world.[113-115]

The Ras proteins are "G proteins", proteins that bind GTP and have endogenous GTPase activity capable of converting GTP to GDP. Like other G proteins in their inactive form, Ras proteins bind GDP. When activated by extracellular signals transduced through membrane receptors, the activated G protein binds GTP and displaces GDP. In their active form, Ras proteins are able to interact with Raf, a serine-threonine protein kinase, to activate MAP kinases which in turn carry the signal further downstream.[107,116] The intrinsic GTPase activity of G proteins converts GTP to GDP with concomitant reversion to the inactivated state. The GTPase activity of Ras proteins is enhanced by the association of Ras with the GTPase activating protein (GAP); thus, association with GAP is a negative regulatory event for normal Ras proteins.

There are four types of *ras* genes encoding proteins of 21 kDa known as H-, K-, N- and R-*ras*. Activation of H-, K-, and N-*ras* involve mutations at codons 12, 13 and 61 and have been found in association with a variety of human cancers and in preleukemia.[106,117] These mutations appear to cluster around the guanine nucleotide binding cleft according to the crystal structure of H-Ras. Mutations at these codons also result in decreased GTPase activity which in part might be due to the inability of Ras to bind GAP in vivo.

2.2.5 Cytoplasmic Serine/Threonine Kinases

The murine sarcoma virus 3611 encodes v-Raf, whose normal homolog, c-Raf, is a serine-threonine kinase that appears to be critical for early events in the transduction of signals from certain receptors.[118-120] Raf is now known to be activated by direct association with activated Ras at the plasma membrane.[116,121-123] In fact, Raf engineered to be targeted to the plasma membrane is able to transform cells.[124,125] Raf in turn is able to activate MAP kinase through MAP kinase kinase, which has neoplastic transforming activity in tissue culture assays.[126-129] Activation of MAP kinase results in the phosphorylation of a number of transcription factors.[127,130]

The Moloney murine sarcoma virus carries the v-*mos* oncogene that encodes another intriguing protein Ser/Thr-kinase.[131-133] The Mos protein participates in the regulation of cyclin dependent kinases. In fact Mos associates with $p34^{cdc2}$ to form maturation promoting factor in *Xenopus* to release oocytes arrested in meiosis.[134-137] Although the exact mechanisms by which Mos transforms mammalian cells are unknown, its interaction with tubulin and the cyclin system is likely to be the critical effect.[136,138-140]

2.2.6 Cell Cycle Machinery

A number of cyclins, which are involved in the regulation of cell cycle kinetics, are implicated in oncogenesis. The cyclins are the regulatory subunits of cell cycle complexes containing cyclin-dependent kinases. Cyclin A has been implicated in the genesis of a case of hepatocellular carcinoma in which the hepatitis genome integrated near the cyclin A gene.[141-143] Cyclin D1 (*bcl*-1) or PRAD1 is activated in lymphomas and parathyroid adenoma.[144,145] Bcl-1, which is involved in t(11;14) translocations and is cyclin D1, is altered in lymphomas specifically in essentially all mantle cell lymphomas and therefore is now considered a marker for this disease.[146-152] Moreover, many of the cyclin mRNAs are found to be highly elevated in many human breast carcinomas. A minority of colon carcinoma cases display cyclin gene amplification.[153]

2.2.7 Transcription Factors

Afferent receptor signals percolate through cytoplasmic messengers and terminate in an alteration of transcription factor activities within the cell nucleus. Like all other steps of signal transduction, transcription factors are also targets for mutations that convert normal cellular genes into oncogenes. Among the better studied transcription factor oncogenes are *ets*, *myc*, *myb*, *fos* and *jun*. The v-*myb* oncogene sustains mutations that result in the loss of negative regulatory domains of the protein yielding a transcription factor that is activated constitutively. Fos and Jun are similarly activated. As important in human cancers is the deregulated expression of Myc.

2.3 TUMOR SUPPRESSOR GENES: LOSS OF THE GOOD AND GAIN OF THE BAD AND UGLY

Some of the early evidence for the existence of tumor suppressor genes came from the studies of somatic cell hybrids formed from normal and neoplastic cells. Surprisingly, such hybrids frequently lacked the ability to form tumors suggesting the existence of genes that inhibit tumor formation.[154,155] These early observations led to the identification of genes normally present that lose the ability to suppress tumor formation when both alleles are mutated. The prototypes are the retinoblastoma protein (pRb), pWT1, and p53.

After analyzing the epidemiology of retinoblastoma, Knudson in the 1970s proposed that familial retinoblastoma results from a germline mutation of one of a pair of inherited alleles that must be complemented by a subsequent acquired mutation in the other allele. In 1983, it became clear that the gene mutated in retinoblastoma is the Rb gene. This gene produces a protein, pRb, that is now known to play an important role in the traversal of cells through the G_1/S transition.[156,157] pRb has been shown to associate with proteins involved in cell cycle regulation (cyclin A and D) and transcriptional activation (E2F). The emerging picture for pRb function is a complicated network of protein-protein interactions.

Analysis of human cancers for allelic loss of heterozygosity has become a paradigm for the identification of tumor suppressor genes, such as WT1 and p53 (Fig. 2.4). The study of Wilms tumor led to the identification of a locus termed WT1 that encodes a transcriptional repressor.[158-163] WT1, however, does not account for genetic alterations found in a large subset of Wilms' tumor, suggesting that another Wilms' tumor gene exists on chromosome 11.[159,164-168] Although initially thought to be an oncogene, p53 was later established as a tumor suppressor gene.[155,157,169,170] p53 is a transcriptional regulatory protein that participates in the regulation of cells traversing through the G_1/S transition. p53 was recognized as a tumor suppressor gene in colon cancers due to a frequent specific loss of a portion of chromosome 17p, where the p53 gene is located, and mutations afflicting the remaining p53 allele.[171] In fact, p53 is one of the most frequently mutated genes in human cancers, suggesting that loss of the normal function of tumor suppressor gene products contributes significantly to neoplastic transformation.

2.4 MISMATCH REPAIR GENES

The presence of microsatellite instability in human cancers was thought to reflect a mutator phenotype that results from defective DNA mismatch repair.[172-178] Microsatellites are DNA sequences containing short (1 to 5 bp) nucleotide repeats of fixed length for a specific allele (Fig. 2.5).[178,179] There may be multiple alleles in the population for a specific microsatellite repeat. However, for an individual, two normal alleles derived from each parent are expected. Microsatellite repeats in cancers and normal cells sustain defects during DNA replication. If these defects are not corrected due to defective DNA mismatch repair, then additional acquired alleles are detected in cancer cells. This phenomenon is termed microsatellite instability. Studies of the yeast and bacterial DNA repair system led to the identification of critical repair proteins termed MutS and MutL and their human counterparts. The human homolog of MutS, termed hMSH2, is encoded by a gene located on chromosome 2. This locus was rapidly recognized as

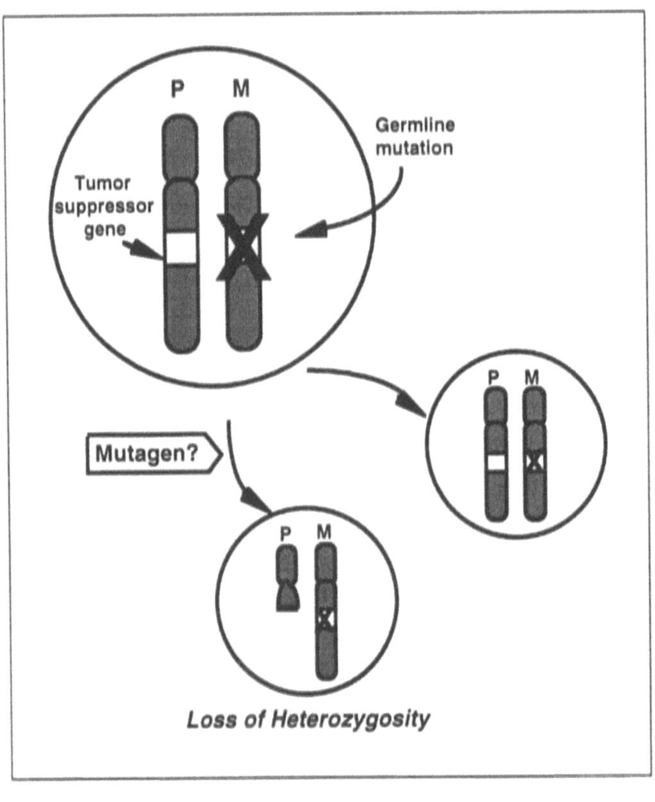

Fig. 2.4. Tumor suppressor genes. The existence of tumor suppressor genes was first elegantly demonstrated by seminal studies of retinoblastoma, which was suspected to be caused by a germline mutation of the retinoblastoma gene. A second hit due to mutagenesis, such as chromosomal deletion, results in loss of the remaining allele (loss of heterozygosity) and tumorigenesis. Somatic mutations of both alleles may also occur by chromosomal deletion, rearrangement, or point mutations resulting in inactivation of tumor suppressor genes. P = paternal allele; M = maternal allele.

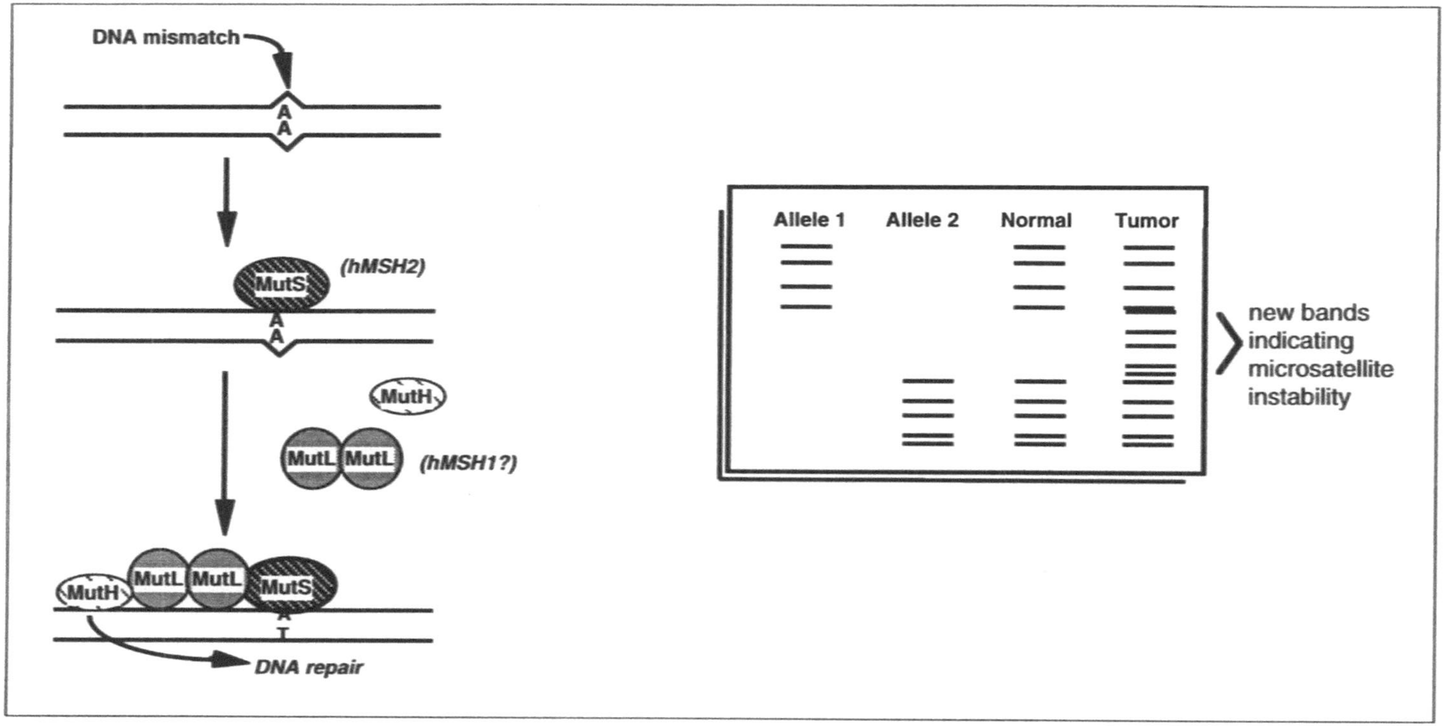

Fig. 2.5. Alteration of the DNA repair machinery in neoplasia. The repair of DNA mismatch has been well studied in bacteria and yeast (depicted by the left diagram). MutS binds to and recognizes DNA mismatch with subsequent assembly of MutL and MutH and repair of the defect. The human MutS homolog, hMSH2, has been found to be mutated in familial non-polyposis colorectal cancer as well as in sporadic colorectal cancers. Defect in mismatch repair is manifested by microsatellite instability (right diagram). Microsatellite repeats of short nucleotide sequences, such as CACACACA, may be polymorphic and appear as distinct ladders of PCR amplified bands. For example, distinct parental alleles (1 and 2) may be detected in normal tissues. With loss of normal mismatch repair, the microsatellites may gain or lose repeats resulting in genome instability and novel patterns of the repeats in tumor cells.

the location of the gene whose loss is responsible for hereditary nonpolyposis colorectal cancer.[172,175,180-188] Out of a cloud of intriguing observations arose a new paradigm for acquired mutations in human cancers: loss of DNA mismatch repair gene function results in the acquisition of additional mutations. Mutations leading to the loss of tumor suppressor genes or activation of oncogenes are selected in the process of tumor development.

2.5 SUFFICIENCY AND NECESSITY OF GENES FOR ONCOGENESIS

The genesis of human cancers involves a myriad of genetic alterations that result in the gain of oncogene function or loss of tumor suppressor gene function (Fig. 2.6). Whether various

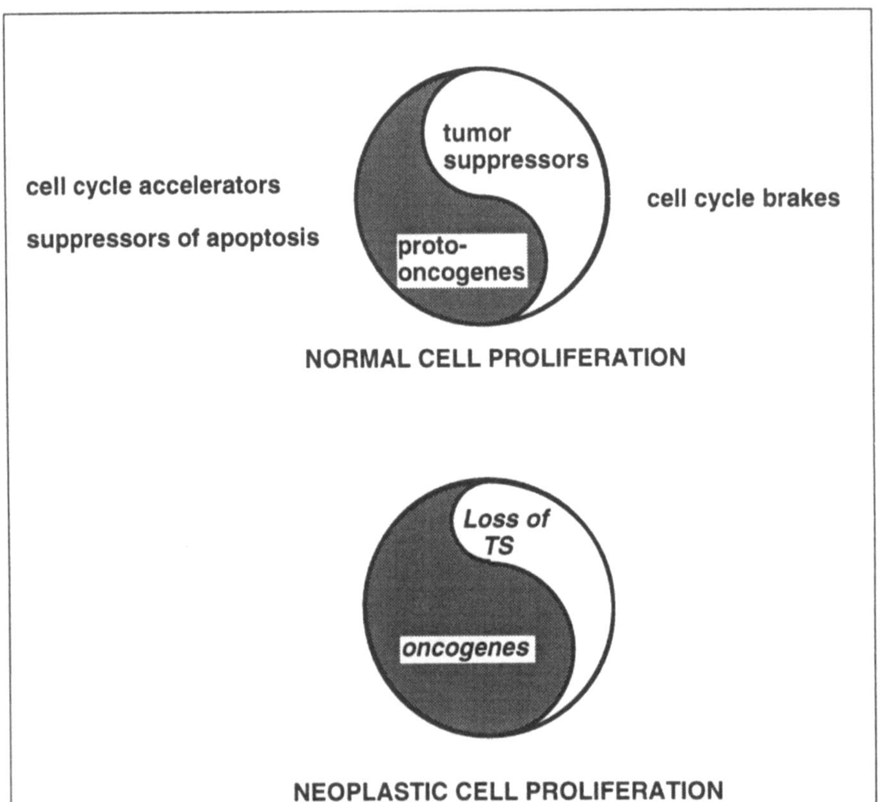

Fig. 2.6. The Zen of neoplastic transformation. Normal cell proliferation results from balanced interactions of proto-oncogenes (accelerators) and tumor suppressor genes (brakes) with the cell cycle machinery. Increased oncogene activity and diminished tumor suppressor (TS) function contribute to neoplastic transformation.

oncogenes or tumor suppressor genes are necessary or in some cases sufficient for inducing the neoplastic phenotype depends heavily on the context of the cell. For example, colon tumorigenesis appears to involve at least six independent events as predicted by mathematical modeling of the age dependent incidence of colon cancer. The work of Volgelstein and coworkers has led to the identification of at least this many genetic alterations in human colon cancer (Fig. 2.7).[117,171,189,190] How many of these events are in fact biologically redundant in individual patients versus the number of events recognized in studying a population of patients is not clear. That is, not all events are necessary for colon tumorigenesis, yet in some individuals several of the many events identified may be sufficient for tumorigenesis. This concept is particularly important to note, since genetic studies using knock-out mice in which specific genetic loci are disrupted in mice have indicated that there are significant redundancies in signal transduction pathways.

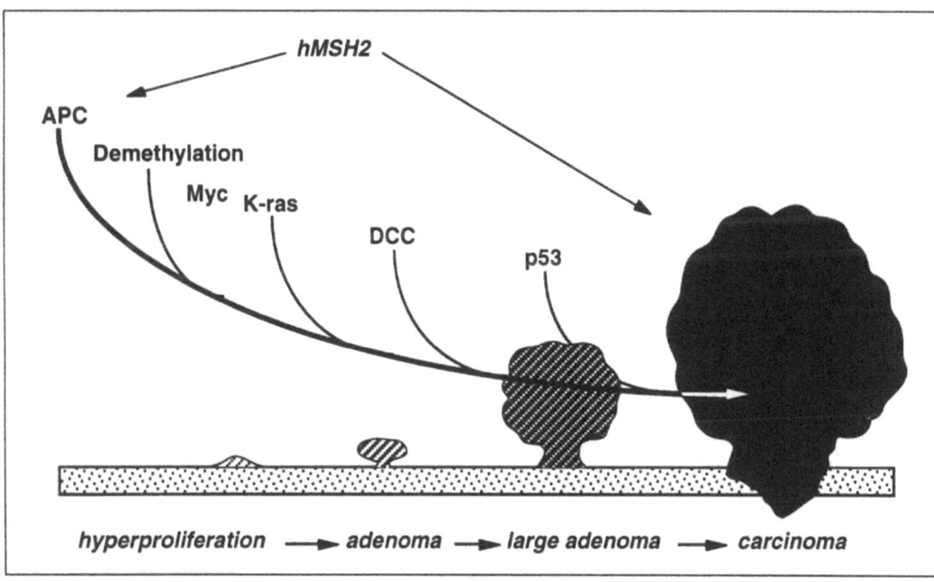

Fig. 2.7. The colorectal cancer model. Colorectal carcinoma results from the accumulation of mutiple genetic alterations. Loss of the familial adenomatous polyposis gene (APC) function is associated with hyperproliferation of intestinal mucosal cells. Adenomas are frequently found to contain mutated K-ras and augmented Myc expression. Loss of DCC and p53 are frequently detected in large adenomas and carcinomas. Defect in DNA mismatch repair (indicated by hMSH2 whose mutation accounts for a large number of hereditary non-polyposis colorectal cancers) is likely to contribute to genomic instability in all stages of colon tumorigenesis.

In contrast to colon tumorigenesis, certain childhood cancers such as bone cancers, retinoblastoma and perhaps leukemias may require fewer genetic alterations to manifest clinically. The retinoblastoma paradigm suggests that a specific human cancer requires mutation of the normal allele in addition to the inherited defect to present as a clinically significant problem. The complementation between oncogenes and tumor suppressor genes is another significant factor in determining oncogenesis. Redundancy in the function of various oncogene products appears to be responsible for the specific complementation between sets of genes in transforming cells in tissue culture. By extrapolation, similar complementation are likely to exist in vivo. Hence, while combinations of genetic alterations might be sufficient for tumorigenesis, any individual alteration might be redundant and not absolutely required for the genesis of specific neoplasms.

References

1. Dyson N, Buchkovich K, Whyte P, Harlow E. The cellular 107K protein that binds to adenovirus E1A also associates with the large T antigens of SV40 and JC virus. Cell 1989; 58:249-55.
2. Dilworth SM. Cell alterations induced by the large T-antigens of SV40 and polyoma virus. [Review]. Semin Cancer Biol 1990; 1:407-14.
3. Butel JS. SV40 large T-antigen: dual oncogene. Cancer Surv 1986; 5:343-65.
4. Lane DP, Simanis V, Bartsch R, Yewdell J, Gannon J, Mole S. Cellular targets for SV40 large T-antigen. Proc R Soc Lond B Biol Sci 1985; 226:25-42.
5. Zhu JY, Abate M, Rice PW, Cole CN. The ability of simian virus 40 large T antigen to immortalize primary mouse embryo fibroblasts cosegregates with its ability to bind to p53. J Virol 1991; 65:6872-80.
6. Ewen ME, Ludlow JW, Marsilio E, et al. An N-terminal transformation-governing sequence of SV40 large T antigen contributes to the binding of both p110Rb and a second cellular protein, p120. Cell 1989; 58:257-67.
7. Tack LC, Wright JH, Deb SP, Tegtmeyer P. The p53 complex from monkey cells modulates the biochemical activities of simian virus 40 large T antigen. J Virol 1989; 63:1310-7.
8. Tan TH, Wallis J, Levine AJ. Identification of the p53 protein domain involved in formation of the simian virus 40 large T-antigen-p53 protein complex. J Virol 1986; 59:574-83.

9. Yew PR, Liu X, Berk AJ. Adenovirus E1B oncoprotein tethers a transcriptional repression domain to p53. Genes Dev 1994; 8:190-202.
10. Lin J, Chen J, Elenbaas B, Levine AJ. Several hydrophobic amino acids in the p53 amino-terminal domain are required for transcriptional activation, binding to mdm-2 and the adenovirus 5 E1B 55-kD protein. Genes Dev 1994; 8:1235-46.
11. White E, Cipriani R, Sabbatini P, Denton A. Adenovirus E1B 19-kilodalton protein overcomes the cytotoxicity of E1A proteins. J Virol 1991; 65:2968-78.
12. Sabbatini P, Chiou SK, Rao L, White E. Modulation of p53-mediated transcriptional repression and apoptosis by the adenovirus E1B 19K protein. Mol Cell Biol 1995; 15:1060-70.
13. Boyd JM, Malstrom S, Subramanian T, et al. Adenovirus E1B 19 kDa and Bcl-2 proteins interact with a common set of cellular proteins. Cell 1994; 79:1121.
14. Boyd JM, Malstrom S, Subramanian T, et al. Adenovirus E1B 19 kDa and Bcl-2 proteins interact with a common set of cellular proteins [see comments] [published erratum appears in Cell 1994 Dec 16;79(6):following 1120]. Cell 1994; 79:341-51.
15. Farrow SN, White JH, Martinou I, et al. Cloning of a bcl-2 homologue by interaction with adenovirus E1B 19K. Nature 1995; 374:731-3.
16. Scheffner M, Huibregtse JM, Vierstra RD, Howley PM. The HPV-16 E6 and E6-AP complex functions as a ubiquitin-protein ligase in the ubiquitination of p53. Cell 1993; 75:495-505.
17. Ciechanover A, Shkedy D, Oren M, Bercovich B. Degradation of the tumor suppressor protein p53 by the ubiquitin-mediated proteolytic system requires a novel species of ubiquitin-carrier protein, E2. J Biol Chem 1994; 269:9582-9.
18. Rolfe M, Beer-Romero P, Glass S, et al. Reconstitution of p53-ubiquitinylation reactions from purified components: the role of human ubiquitin-conjugating enzyme UBC4 and E6-associated protein (E6AP). Proc Natl Acad Sci USA 1995; 92:3264-8.
19. Davies R, Hicks R, Crook T, Morris J, Vousden K. Human papillomavirus type 16 E7 associates with a histone H1 kinase and with p107 through sequences necessary for transformation. J Virol 1993; 67:2521-8.
20. Arroyo M, Bagchi S, Raychaudhuri P. Association of the human papillomavirus type 16 E7 protein with the S-phase-specific E2F-cyclin A complex. Mol Cell Biol 1993; 13:6537-46.
21. Ciccolini F, Di Pasquale G, Carlotti F, Crawford L, Tommasino M. Functional studies of E7 proteins from different HPV types. Oncogene 1994; 9:2633-8.

22. Seto E, Mitchell PJ, Yen TS. Transactivation by the hepatitis B virus X protein depends on AP-2 and other transcription factors. Nature 1990; 344:72-4.
23. Balsano C, Avantaggiati ML, Natoli G, et al. Full-length and truncated versions of the hepatitis B virus (HBV) X protein (pX) transactivate the cmyc protooncogene at the transcriptional level. Biochem Biophys Res Commun 1991; 176:985-92.
24. Bishop JM. Retroviruses. [Review]. Annu Rev Biochem 1978; 47:35-88.
25. Bishop JM. Retroviruses and cancer genes. [Review]. Adv Cancer Res 1982; 37:1-32.
26. Bishop JM. Viruses, genes, and cancer. II. Retroviruses and cancer genes. Cancer 1985; 55:2329-33.
27. Swanstrom R, Parker RC, Varmus HE, Bishop JM. Transduction of a cellular oncogene: the genesis of Rous sarcoma virus. Proc Natl Acad Sci USA 1983; 80:2519-23.
28. Bister K, Jansen HW. Oncogenes in retroviruses and cells: biochemistry and molecular genetics. [Review]. Adv Cancer Res 1986; 47:99-188.
29. Hayward WS, Neel BG, Astrin SM. Activation of a cellular onc gene by promoter insertion in ALV-induced lymphoid leukosis. Nature 1981; 290:475-80.
30. Neel BG, Hayward WS, Robinson HL, Fang J, Astrin SM. Avian leukosis virus-induced tumors have common proviral integration sites and synthesize discrete new RNAs: oncogenesis by promoter insertion. Cell 1981; 23:323-34.
31. Dang CV. Oncogenes and proto-oncogenes: General concepts. In: Benz C, Liu E, eds. Oncogenes: Kluwer Academic Publishers, 1989:3-24.
32. Alitalo K, Schwab M. Oncogene amplification in tumor cells. [Review]. Adv Cancer Res 1986; 47:235-81.
33. Bishop JM. Molecular themes in oncogenesis. [Review]. Cell 1991; 64:235-48.
34. Bishop JM. Cellular oncogenes and retroviruses. [Review]. Annu Rev Biochem 1983; 52:301-54.
35. Doolittle RF, Hunkapiller MW, Hood LE, et al. Simian sarcoma virus onc gene, v-sis, is derived from the gene (or genes) encoding a platelet-derived growth factor. Science 1983; 221:275-7.
36. Keating MT, Williams LT. Autocrine stimulation of intracellular PDGF receptors in v-sis-transformed cells. Science 1988; 239:914-6.
37. Cavalieri F, Goldfarb M. Growth factor-deprived BALB/c 3T3 murine fibroblasts can enter the S phase after induction of c-myc gene expression. Mol Cell Biol 1987; 7:3554-60.
38. Zhan X, Goldfarb M. Growth factor requirements of oncogene-transformed NIH 3T3 and BALB/c 3T3 cells cultured in defined media. Mol Cell Biol 1986; 6:3541-4.

39. Fantl WJ, Escobedo JA, Martin GA, et al. Distinct phosphotyrosines on a growth factor receptor bind to specific molecules that mediate different signaling pathways. Cell 1992; 69:413-23.
40. Nusse R, van Ooyen A, Cox D, Fung YK, Varmus H. Mode of proviral activation of a putative mammary oncogene (int-1) on mouse chromosome 15. Nature 1984; 307:131-6.
41. van Ooyen A, Nusse R. Structure and nucleotide sequence of the putative mammary oncogene int-1; proviral insertions leave the protein-encoding domain intact. Cell 1984; 39:233-40.
42. Hing HK, Sun X, Artavanis-Tsakonas S. Modulation of wingless signaling by Notch in Drosophila. Mech Dev 1994; 47:261-8.
43. Siegfried E, Wilder EL, Perrimon N. Components of wingless signalling in Drosophila. Nature 1994; 367:76-80.
44. Papkoff J, Schryver B. Secreted int-1 protein is associated with the cell surface. Mol Cell Biol 1990; 10:2723-30.
45. Bradley RS, Brown AM. The proto-oncogene int-1 encodes a secreted protein associated with the extracellular matrix. EMBO J 1990; 9:1569-75.
46. Hinck L, Nathke IS, Papkoff J, Nelson WJ. Beta-catenin: a common target for the regulation of cell adhesion by Wnt-1 and Src signaling pathways. [Review]. Trends Biochem Sci 1994; 19:538-42.
47. Dickinson ME, Krumlauf R, McMahon AP. Evidence for a mitogenic effect of Wnt-1 in the developing mammalian central nervous system. Development 1994; 120:1453-71.
48. Burrus LW. Wnt-1 as a short-range signaling molecule. [Review]. Bioessays 1994; 16:155-7.
49. Hinck L, Nelson WJ, Papkoff J. Wnt-1 modulates cell-cell adhesion in mammalian cells by stabilizing beta-catenin binding to the cell adhesion protein cadherin. J Cell Biol 1994; 124:729-41.
50. Smolich BD, McMahon JA, McMahon AP, Papkoff J. Wnt family proteins are secreted and associated with the cell surface. Mol Biol Cell 1993; 4:1267-75.
51. Parkin NT, Kitajewski J, Varmus HE. Activity of Wnt-1 as a transmembrane protein. Genes Dev 1993; 7:2181-93.
52. Wellstein A, Lupu R, Zugmaier G, et al. Autocrine growth stimulation by secreted Kaposi fibroblast growth factor but not by endogenous basic fibroblast growth factor. Cell Growth Differ 1990; 1:63-71.
53. Delli Bovi P, Curatola AM, Kern FG, Greco A, Ittmann M, Basilico C. An oncogene isolated by transfection of Kaposi's sarcoma DNA encodes a growth factor that is a member of the FGF family. Cell 1987; 50:729-37.
54. Taira M, Yoshida T, Miyagawa K, Sakamoto H, Terada M, Sugimura T. cDNA sequence of human transforming gene hst and identification of the coding sequence required for transforming activity. Proc Natl Acad Sci USA 1987; 84:2980-4.

55. Peters G, Brookes S, Smith R, Placzek M, Dickson C. The mouse homolog of the hst/k-FGF gene is adjacent to int-2 and is activated by proviral insertion in some virally induced mammary tumors. Proc Natl Acad Sci USA 1989; 86:5678-82.
56. Sakamoto H, Mori M, Taira M, et al. Transforming gene from human stomach cancers and a noncancerous portion of stomach mucosa. Proc Natl Acad Sci USA 1986; 83:3997-4001.
57. Mansour SL, Martin GR. Four classes of mRNA are expressed from the mouse int-2 gene, a member of the FGF gene family. EMBO J 1988; 7:2035-41.
58. Downward J, Yarden Y, Mayes E, et al. Close similarity of epidermal growth factor receptor and v-erb-B oncogene protein sequences. Nature 1984; 307:521-7.
59. Schechter AL, Stern DF, Vaidyanathan L, et al. The neu oncogene: an erb-B-related gene encoding a 185,000-Mr tumour antigen. Nature 1984; 312:513-6.
60. Bargmann CI, Hung MC, Weinberg RA. The neu oncogene encodes an epidermal growth factor receptor-related protein. Nature 1986; 319:226-30.
61. Schechter AL, Hung MC, Vaidyanathan L, et al. The neu gene: an erbB-homologous gene distinct from and unlinked to the gene encoding the EGF receptor. Science 1985; 229:976-8.
62. Slamon DJ, Clark GM, Wong SG, Levin WJ, Ullrich A, McGuire WL. Human breast cancer: correlation of relapse and survival with amplification of the HER-2/neu oncogene. Science 1987; 235:177-82.
63. Borg A, Baldetorp B, Ferno M, Killander D, Olsson H, Sigurdsson H. ERBB2 amplification in breast cancer with a high rate of proliferation. Oncogene 1991; 6:137-43.
64. Kallioniemi OP, Kallioniemi A, Kurisu W, et al. ERBB2 amplification in breast cancer analyzed by fluorescence in situ hybridization. Proc Natl Acad Sci USA 1992; 89:5321-5.
65. Walker RA, Varley JM. The molecular pathology of human breast cancer. [Review]. Cancer Surv 1993; 16:31-57.
66. Holmes WE, Sliwkowski MX, Akita RW, et al. Identification of heregulin, a specific activator of p185erbB2. Science 1992; 256:1205-10.
67. Peles E, Yarden Y. Neu and its ligands: from an oncogene to neural factors. [Review]. Bioessays 1993; 15:815-24.
68. Plowman GD, Green JM, Culouscou JM, Carlton GW, Rothwell VM, Buckley S. Heregulin induces tyrosine phosphorylation of HER4/p180erbB4. Nature 1993; 366:473-5.
69. Rettenmier CW, Roussel MF, Quinn CO, Kitchingman GR, Look AT, Sherr CJ. Transmembrane orientation of glycoproteins encoded by the v-fms oncogene. Cell 1985; 40:971-81.
70. Sacca R, Stanley ER, Sherr CJ, Rettenmier CW. Specific binding of the mononuclear phagocyte colony-stimulating factor CSF-1 to the product of the v-fms oncogene. Proc Natl Acad Sci USA 1986; 83:3331-5.

71. Wheeler EF, Rettenmier CW, Look AT, Sherr CJ. The v-fms oncogene induces factor independence and tumorigenicity in CSF-1 dependent macrophage cell line. Nature 1986; 324:377-80.
72. Roussel MF, Dull TJ, Rettenmier CW, Ralph P, Ullrich A, Sherr CJ. Transforming potential of the c-fms proto-oncogene (CSF-1 receptor). Nature 1987; 325:549-52.
73. Sherr CJ. The fms oncogene. [Review]. Biochim Biophys Acta 1988; 948:225-43.
74. Copeland NG, Gilbert DJ, Cho BC, et al. Mast cell growth factor maps near the steel locus on mouse chromosome 10 and is deleted in a number of steel alleles. Cell 1990; 63:175-83.
75. Anderson DM, Lyman SD, Baird A, et al. Molecular cloning of mast cell growth factor, a hematopoietin that is active in both membrane bound and soluble forms [published erratum appears in Cell 1990 Nov 30;63(5):following 1112]. Cell 1990; 63:235-43.
76. Huang E, Nocka K, Beier DR, et al. The hematopoietic growth factor KL is encoded by the Sl locus and is the ligand of the c-kit receptor, the gene product of the W locus. Cell 1990; 63:225-33.
77. Matsui Y, Zsebo KM, Hogan BL. Embryonic expression of a haematopoietic growth factor encoded by the Sl locus and the ligand for c-kit. Nature 1990; 347:667-9.
78. Flanagan JG, Leder P. The kit ligand: a cell surface molecule altered in steel mutant fibroblasts. Cell 1990; 63:185-94.
79. Parker RC, Varmus HE, Bishop JM. Cellular homologue (c-src) of the transforming gene of Rous sarcoma virus: isolation, mapping, and transcriptional analysis of c-src and flanking regions. Proc Natl Acad Sci USA 1981; 78:5842-6.
80. Cohen GB, Ren R, Baltimore D. Modular binding domains in signal transduction proteins. [Review]. Cell 1995; 80:237-48.
81. Heisterkamp N, Stam K, Groffen J, de Klein A, Grosveld G. Structural organization of the bcr gene and its role in the Ph' translocation. Nature 1985; 315:758-61.
82. Shtivelman E, Lifshitz B, Gale RP, Roe BA, Canaani E. Alternative splicing of RNAs transcribed from the human abl gene and from the bcr-abl fused gene. Cell 1986; 47:277-84.
83. Ben-Neriah Y, Daley GQ, Mes-Masson AM, Witte ON, Baltimore D. The chronic myelogenous leukemia-specific P210 protein is the product of the bcr/abl hybrid gene. Science 1986; 233:212-4.
84. Collins S, Coleman H, Groudine M. Expression of bcr and bcr-abl fusion transcripts in normal and leukemic cells. Mol Cell Biol 1987; 7:2870-6.
85. McLaughlin J, Chianese E, Witte ON. In vitro transformation of immature hematopoietic cells by the P210 BCR/ABL oncogene product of the Philadelphia chromosome. Proc Natl Acad Sci USA 1987; 84:6558-62.

86. Hariharan IK, Adams JM. cDNA sequence for human bcr, the gene that translocates to the abl oncogene in chronic myeloid leukaemia. EMBO J 1987; 6:115-9.
87. Fainstein E, Marcelle C, Rosner A, et al. A new fused transcript in Philadelphia chromosome positive acute lymphocytic leukaemia. Nature 1987; 330:386-8.
88. Chan LC, Karhi KK, Rayter SI, et al. A novel abl protein expressed in Philadelphia chromosome positive acute lymphoblastic leukaemia. Nature 1987; 325:635-7.
89. Kurzrock R, Shtalrid M, Romero P, et al. A novel c-abl protein product in Philadelphia-positive acute lymphoblastic leukaemia. Nature 1987; 325:631-5.
90. Clark SS, McLaughlin J, Crist WM, Champlin R, Witte ON. Unique forms of the abl tyrosine kinase distinguish Ph1-positive CML from Ph1-positive ALL. Science 1987; 235:85-8.
91. Walker LC, Ganesan TS, Dhut S, et al. Novel chimaeric protein expressed in Philadelphia positive acute lymphoblastic leukaemia. Nature 1987; 329:851-3.
92. Naldini L, Stacchini A, Cirillo DM, Aglietta M, Gavosto F, Comoglio PM. Phosphotyrosine antibodies identify the p210c-abl tyrosine kinase and proteins phosphorylated on tyrosine in human chronic myelogenous leukemia cells. Mol Cell Biol 1986; 6:1803-11.
93. Clark SS, McLaughlin J, Timmons M, et al. Expression of a distinctive BCR-ABL oncogene in Ph1-positive acute lymphocytic leukemia (ALL). Science 1988; 239:775-7.
94. Timmons MS, Witte ON. Structural characterization of the BCR gene product. Oncogene 1989; 4:559-67.
95. Elefanty AG, Hariharan IK, Cory S. bcr-abl, the hallmark of chronic myeloid leukaemia in man, induces multiple haemopoietic neoplasms in mice. Embo J 1990; 9:1069-78.
96. Afar DE, Goga A, McLaughlin J, Witte ON, Sawyers CL. Differential complementation of Bcr-Abl point mutants with c-Myc. Science 1994; 264:424-6.
97. Pendergast AM, Quilliam LA, Cripe LD, et al. BCR-ABL-induced oncogenesis is mediated by direct interaction with the SH2 domain of the GRB-2 adaptor protein. Cell 1993; 75:175-85.
98. Taparowsky E, Suard Y, Fasano O, Shimizu K, Goldfarb M, Wigler M. Activation of the T24 bladder carcinoma transforming gene is linked to a single amino acid change. Nature 1982; 300:762-5.
99. Taparowsky E, Shimizu K, Goldfarb M, Wigler M. Structure and activation of the human N-ras gene. Cell 1983; 34:581-6.
100. Fasano O, Taparowsky E, Fiddes J, Wigler M, Goldfarb M. Sequence and structure of the coding region of the human H-ras-1 gene from T24 bladder carcinoma cells. J Mol Appl Genet 1983; 2:173-80.
101. Der CJ. The ras family of oncogenes. [Review]. Cancer Treat Res 1989; 47:73-119.

102. Casey PJ, Solski PA, Der CJ, Buss JE. p21ras is modified by a farnesyl isoprenoid. Proc Natl Acad Sci USA 1989; 86:8323-7.
103. Buss JE, Solski PA, Schaeffer JP, MacDonald MJ, Der CJ. Activation of the cellular proto-oncogene product p21Ras by addition of a myristylation signal. Science 1989; 243:1600-3.
104. Finkel T, Der CJ, Cooper GM. Activation of ras genes in human tumors does not affect localization, modification, or nucleotide binding properties of p21. Cell 1984; 37:151-8.
105. Der CJ, Finkel T, Cooper GM. Biological and biochemical properties of human rasH genes mutated at codon 61. Cell 1986; 44:167-76.
106. Bos JL, Fearon ER, Hamilton SR, et al. Prevalence of ras gene mutations in human colorectal cancers. Nature 1987; 327:293-7.
107. Maruta H, Burgess AW. Regulation of the Ras signalling network. [Review]. Bioessays 1994; 16:489-96.
108. Schlaepfer DD, Hanks SK, Hunter T, van der Geer P. Integrin-mediated signal transduction linked to Ras pathway by GRB2 binding to focal adhesion kinase. Nature 1994; 372:786-91.
109. Chardin P, Camonis JH, Gale NW, et al. Human Sos1: a guanine nucleotide exchange factor for Ras that binds to GRB2. Science 1993; 260:1338-43.
110. Skolnik EY, Batzer A, Li N, et al. The function of GRB2 in linking the insulin receptor to Ras signaling pathways. Science 1993; 260:1953-5.
111. Buday L, Downward J. Epidermal growth factor regulates p21ras through the formation of a complex of receptor, Grb2 adapter protein, and Sos nucleotide exchange factor. Cell 1993; 73:611-20.
112. Rozakis-Adcock M, Fernley R, Wade J, Pawson T, Bowtell D. The SH2 and SH3 domains of mammalian Grb2 couple the EGF receptor to the Ras activator Sos1. Nature 1993; 363:83-5.
113. Olivier JP, Raabe T, Henkemeyer M, et al. A Drosophila SH2-SH3 adaptor protein implicated in coupling the sevenless tyrosine kinase to an activator of Ras guanine nucleotide exchange, Sos. Cell 1993; 73:179-91.
114. Baltensperger K, Kozma LM, Cherniack AD, et al. Binding of the Ras activator son of sevenless to insulin receptor substrate-1 signaling complexes. Science 1993; 260:1950-2.
115. Karlovich CA, Bonfini L, McCollam L, et al. In vivo functional analysis of the Ras exchange factor son of sevenless. Science 1995; 268:576-9.
116. Avruch J, Zhang XF, Kyriakis JM. Raf meets Ras: completing the framework of a signal transduction pathway. [Review]. Trends Biochem Sci 1994; 19:279-83.
117. Vogelstein B, Fearon ER, Hamilton SR, et al. Genetic alterations during colorectal-tumor development. N Engl J Med 1988; 319:525-32.

118. Sutrave P, Bonner TI, Rapp UR, Jansen HW, Patschinsky T, Bister K. Nucleotide sequence of avian retroviral oncogene v-mil: homologue of murine retroviral oncogene v-raf. Nature 1984; 309:85-8.
119. Klinken SP, Alexander WS, Adams JM. Hemopoietic lineage switch: v-raf oncogene converts Emu-myc transgenic B cells into macrophages. Cell 1988; 53:857-67.
120. Magnuson NS, Beck T, Vahidi H, Hahn H, Smola U, Rapp UR. The Raf-1 serine/threonine protein kinase. [Review]. Semin Cancer Biol 1994; 5:247-53.
121. Fabian JR, Vojtek AB, Cooper JA, Morrison DK. A single amino acid change in Raf-1 inhibits Ras binding and alters Raf-1 function. Proc Natl Acad Sci USA 1994; 91:5982-6.
122. Pumiglia K, Chow YH, Fabian J, Morrison D, Decker S, Jove R. Raf-1 N-terminal sequences necessary for Ras-Raf interaction and signal transduction. Mol Cell Biol 1995; 15:398-406.
123. Koide H, Satoh T, Nakafuku M, Kaziro Y. GTP-dependent association of Raf-1 with Ha-Ras: identification of Raf as a target downstream of Ras in mammalian cells. Proc Natl Acad Sci USA 1993; 90:8683-6.
124. Leevers SJ, Paterson HF, Marshall CJ. Requirement for Ras in Raf activation is overcome by targeting Raf to the plasma membrane. Nature 1994; 369:411-4.
125. Stokoe D, Macdonald SG, Cadwallader K, Symons M, Hancock JF. Activation of Raf as a result of recruitment to the plasma membrane [see comments] [published erratum appears in Science 1994 Dec 16;266(5192):1792-3]. Science 1994; 264:1463-7.
126. Moodie SA, Willumsen BM, Weber MJ, Wolfman A. Complexes of Ras.GTP with Raf-1 and mitogen-activated protein kinase kinase. Science 1993; 260:1658-61.
127. Howe LR, Leevers SJ, Gomez N, Nakielny S, Cohen P, Marshall CJ. Activation of the MAP kinase pathway by the protein kinase raf. Cell 1992; 71:335-42.
128. Kyriakis JM, App H, Zhang XF, et al. Raf-1 activates MAP kinase-kinase. Nature 1992; 358:417-21.
129. Mansour SJ, Matten WT, Hermann AS, et al. Transformation of mammalian cells by constitutively active MAP kinase kinase. Science 1994; 265:966-70.
130. Ruderman JV. MAP kinase and the activation of quiescent cells. [Review]. Curr Opin Cell Biol 1993; 5:207-13.
131. Bold RJ, Donoghue DJ. Biologically active mutants with deletions in the v-mos oncogene assayed with retroviral vectors. Mol Cell Biol 1985; 5:3131-8.
132. Bold RJ, Hannink M, Donoghue DJ. Functions of the mos oncogene family and associated gene products. Cancer Surv 1986; 5:243-55.

133. Freeman RS, Kanki JP, Ballantyne SM, Pickham KM, Donoghue DJ. Effects of the v-mos oncogene on Xenopus development: meiotic induction in oocytes and mitotic arrest in cleaving embryos. J Cell Biol 1990; 111:533-41.
134. Yew N, Mellini ML, Vande Woude GF. Meiotic initiation by the mos protein in Xenopus. Nature 1992; 355:649-52.
135. Daar I, Paules RS, Vande Woude GF. A characterization of cytostatic factor activity from Xenopus eggs and c-mos-transformed cells. J Cell Biol 1991; 114:329-35.
136. Zhou RP, Oskarsson M, Paules RS, Schulz N, Cleveland D, Vande Woude GF. Ability of the c-mos product to associate with and phosphorylate tubulin. Science 1991; 251:671-5.
137. Yew N, Oskarsson M, Daar I, Blair DG, Vande Woude GF. mos gene transforming efficiencies correlate with oocyte maturation and cytostatic factor activities. Mol Cell Biol 1991; 11:604-10.
138. Fukasawa K, Vande Woude GF. Mos overexpression in Swiss 3T3 cells induces meiotic-like alterations of the mitotic spindle. Proc Natl Acad Sci USA 1995; 92:3430-4.
139. Wang XM, Yew N, Peloquin JG, Vande Woude GF, Borisy GG. Mos oncogene product associates with kinetochores in mammalian somatic cells and disrupts mitotic progression. Proc Natl Acad Sci USA 1994; 91:8329-33.
140. Yew N, Strobel M, Vande Woude GF. Mos and the cell cycle: the molecular basis of the transformed phenotype. [Review]. Curr Opin Genet Dev 1993; 3:19-25.
141. Brechot C. Oncogenic activation of cyclin A. [Review]. Curr Opin Genet Dev 1993; 3:11-8.
142. Henglein B, Chenivesse X, Wang J, Eick D, Brechot C. Structure and cell cycle-regulated transcription of the human cyclin A gene. Proc Natl Acad Sci USA 1994; 91:5490-4.
143. Wang J, Zindy F, Chenivesse X, Lamas E, Henglein B, Brechot C. Modification of cyclin A expression by hepatitis B virus DNA integration in a hepatocellular carcinoma. Oncogene 1992; 7:1653-6.
144. Rosenberg CL, Motokura T, Kronenberg HM, Arnold A. Coding sequence of the overexpressed transcript of the putative oncogene PRAD1/cyclin D1 in two primary human tumors. Oncogene 1993; 8:519-21.
145. Motokura T, Arnold A. Cyclin D and oncogenesis. [Review]. Curr Opin Genet Dev 1993; 3:5-10.
146. Bosch F, Jares P, Campo E, et al. PRAD-1/cyclin D1 gene overexpression in chronic lymphoproliferative disorders: a highly specific marker of mantle cell lymphoma. Blood 1994; 84:2726-32.
147. de Boer CJ, Loyson S, Kluin PM, Kluin-Nelemans HC, Schuuring E, van Krieken JH. Multiple breakpoints within the BCL-1 locus in B-cell lymphoma: rearrangements of the cyclin D1 gene. Cancer Res 1993; 53:4148-52.

148. Wotherspoon AC, Pan LX, Diss TC, Isaacson PG. A genotypic study of low grade B-cell lymphomas, including lymphomas of mucosa associated lymphoid tissue (MALT). J Pathol 1990; 162:135-40.
149. Ince C, Blick M, Lee M, et al. Bcl-1 gene rearrangements in B cell lymphoma. Leukemia 1988; 2:343-6.
150. Tsujimoto Y, Jaffe E, Cossman J, Gorham J, Nowell PC, Croce CM. Clustering of breakpoints on chromosome 11 in human B-cell neoplasms with the t(11;14) chromosome translocation. Nature 1985; 315:340-3.
151. Erikson J, Finan J, Tsujimoto Y, Nowell PC, Croce CM. The chromosome 14 breakpoint in neoplastic B cells with the t(11;14) translocation involves the immunoglobulin heavy chain locus. Proc Natl Acad Sci USA 1984; 81:4144-8.
152. Withers DA, Harvey RC, Faust JB, Melnyk O, Carey K, Meeker TC. Characterization of a candidate bcl-1 gene. Mol Cell Biol 1991; 11:4846-53.
153. Leach FS, Elledge SJ, Sherr CJ, et al. Amplification of cyclin genes in colorectal carcinomas. Cancer Res 1993; 53:1986-9.
154. Marshall CJ. Tumor suppressor genes. [Review]. Cell 1991; 64:313-26.
155. Levine AJ, Momand J. Tumor suppressor genes: the p53 and retinoblastoma sensitivity genes and gene products. [Review]. Biochim Biophys Acta 1990; 1032:119-36.
156. Nevins JR. E2F: a link between the Rb tumor suppressor protein and viral oncoproteins. Science 1992; 258:424-9.
157. Levine AJ. Tumor suppressor genes. [Review]. Bioessays 1990; 12:60-6.
158. Haber DA, Park S, Maheswaran S, et al. WT1-mediated growth suppression of Wilms tumor cells expressing a WT1 splicing variant. Science 1993; 262:2057-9.
159. Coppes MJ, Campbell CE, Williams BR. The role of WT1 in Wilms tumorigenesis. [Review]. FASEB J 1993; 7:886-95.
160. Varanasi R, Bardeesy N, Ghahremani M, et al. Fine structure analysis of the WT1 gene in sporadic Wilms tumors. Proc Natl Acad Sci USA 1994; 91:3554-8.
161. Pelletier J, Bruening W, Kashtan CE, et al. Germline mutations in the Wilms' tumor suppressor gene are associated with abnormal urogenital development in Denys-Drash syndrome. Cell 1991; 67:437-47.
162. Haber DA, Buckler AJ, Glaser T, et al. An internal deletion within an 11p13 zinc finger gene contributes to the development of Wilms' tumor. Cell 1990; 61:1257-69.
163. Madden SL, Cook DM, Morris JF, Gashler A, Sukhatme VP, Rauscher F, III. Transcriptional repression mediated by the WT1 Wilms tumor gene product. Science 1991; 253:1550-3.
164. Gessler M, Konig A, Arden K, et al. Infrequent mutation of the WT1 gene in 77 Wilms' Tumors. Hum Mutat 1994; 3:212-22.
165. Brown KW, Wilmore HP, Watson JE, Mott MG, Berry PJ, Maitland NJ. Low frequency of mutations in the WT1 coding region in Wilms' tumor. Genes Chromosom Cancer 1993; 8:74-9.

166. Haber DA, Housman DE. Role of the WT1 gene in Wilms' tumour. [Review]. Cancer Surv 1992; 12:105-17.
167. Huff V, Saunders GF. Wilms tumor genes. [Review]. Biochim Biophys Acta 1993; 1155:295-306.
168. Hastie ND. Wilms' tumour gene and function. [Review]. Curr Opin Genet Dev 1993; 3:408-13.
169. Finlay CA, Hinds PW, Levine AJ. The p53 proto-oncogene can act as a suppressor of transformation. Cell 1989; 57:1083-93.
170. Hinds P, Finlay C, Levine AJ. Mutation is required to activate the p53 gene for cooperation with the ras oncogene and transformation. J Virol 1989; 63:739-46.
171. Fearon ER, Vogelstein B. A genetic model for colorectal tumorigenesis. [Review]. Cell 1990; 61:759-67.
172. Modrich P. Mismatch repair, genetic stability, and cancer. [Review]. Science 1994; 266:1959-60.
173. Merlo A, Mabry M, Gabrielson E, Vollmer R, Baylin SB, Sidransky D. Frequent microsatellite instability in primary small cell lung cancer. Cancer Res 1994; 54:2098-101.
174. Horii A, Han HJ, Shimada M, et al. Frequent replication errors at microsatellite loci in tumors of patients with multiple primary cancers. Cancer Res 1994; 54:3373-5.
175. Jiricny J. Colon cancer and DNA repair: have mismatches met their match?. [Review]. Trends Genet 1994; 10:164-8.
176. Orth K, Hung J, Gazdar A, Bowcock A, Mathis JM, Sambrook J. Genetic instability in human ovarian cancer cell lines. Proc Natl Acad Sci USA 1994; 91:9495-9.
177. Fong KM, Zimmerman PV, Smith PJ. Microsatellite instability and other molecular abnormalities in non-small cell lung cancer. Cancer Res 1995; 55:28-30.
178. Parsons R, Li GM, Longley MJ, et al. Hypermutability and mismatch repair deficiency in RER+ tumor cells. Cell 1993; 75:1227-36.
179. Tautz D, Schlotterer. Simple sequences. [Review]. Curr Opin Genet Dev 1994; 4:832-7.
180. Liu B, Nicolaides NC, Markowitz S, et al. Mismatch repair gene defects in sporadic colorectal cancers with microsatellite instability. Nat Genet 1995; 9:48-55.
181. Nystrom-Lahti M, Parsons R, Sistonen P, et al. Mismatch repair genes on chromosomes 2p and 3p account for a major share of hereditary nonpolyposis colorectal cancer families evaluable by linkage. Am J Hum Genet 1994; 55:659-65.
182. Bronner CE, Baker SM, Morrison PT, et al. Mutation in the DNA mismatch repair gene homologue hMLH1 is associated with hereditary non-polyposis colon cancer. Nature 1994; 368:258-61.
183. Hemminki A, Peltomaki P, Mecklin JP, et al. Loss of the wild type MLH1 gene is a feature of hereditary nonpolyposis colorectal cancer. Nat Genet 1994; 8:405-10.

184. Kolodner RD, Hall NR, Lipford J, et al. Structure of the human MLH1 locus and analysis of a large hereditary nonpolyposis colorectal carcinoma kindred for mlh1 mutations. Cancer Res 1995; 55:242-8.
185. Nicolaides NC, Papadopoulos N, Liu B, et al. Mutations of two PMS homologues in hereditary nonpolyposis colon cancer. Nature 1994; 371:75-80.
186. Liu B, Parsons RE, Hamilton SR, et al. hMSH2 mutations in hereditary nonpolyposis colorectal cancer kindreds. Cancer Res 1994; 54:4590-4.
187. Papadopoulos N, Nicolaides NC, Wei YF, et al. Mutation of a mutL homolog in hereditary colon cancer [see comments]. Science 1994; 263:1625-9.
188. Leach FS, Nicolaides NC, Papadopoulos N, et al. Mutations of a mutS homolog in hereditary nonpolyposis colorectal cancer. Cell 1993; 75:1215-25.
189. Vogelstein B, Fearon ER, Kern SE, et al. Allelotype of colorectal carcinomas. Science 1989; 244:207-11.
190. Fearon ER, Hamilton SR, Vogelstein B. Clonal analysis of human colorectal tumors. Science 1987; 238:193-7.

CHAPTER 3

Historical Perspectives of *myc* Gene Studies

3.1 RETROVIRUSES BEARING *myc* ONCOGENE

Much of our current understanding of cancer genes originated from classical studies of avian tumors. The very first documented observation by Rous in 1910 suggested that certain avian sarcomas could be transmitted by a filterable infectious agent.[1-4] Subsequently, isolates of the infectious agents were recognized under the electron microscope as membrane bound particles now known as retroviruses. Five of these retroviruses that have been isolated from avian sarcomas and leukemias are known to bear v-*myc* oncogenes.[5,6]

The MC29 subgroup of avian acute leukemia virus consists of strains MC29, CMII, OK10, and MH2; FH3 is a newly characterized v-*myc* retrovirus.[5,7] All strains of the MC29 subgroup induce myelocytomatosis (hence *myc*) or myeloid leukemia, endothelioma, liver and kidney carcinomas, sarcomas and erythroblastosis in chickens and other birds. Tumors are formed rapidly and often are fatal usually within weeks of infection by these retroviruses.

3.2 MOLECULAR CLONING OF V-*myc* ONCOGENES AND IDENTIFICATION OF THE CELLULAR HOMOLOG

The genetic structures of the genomes for MC29, CMII, OK10, MH2 and FH3 have been delineated and found to contain oncogenic v-*myc* alleles (for review see reference 5) (Fig. 3.1). Molecular cloning of the provirus and determination of its nucleotide sequence were independently determined by several groups of researchers. The genome does not contain a full complement of the three replicative genes (*gag*, *pol* and *env*), but instead the v-*myc*

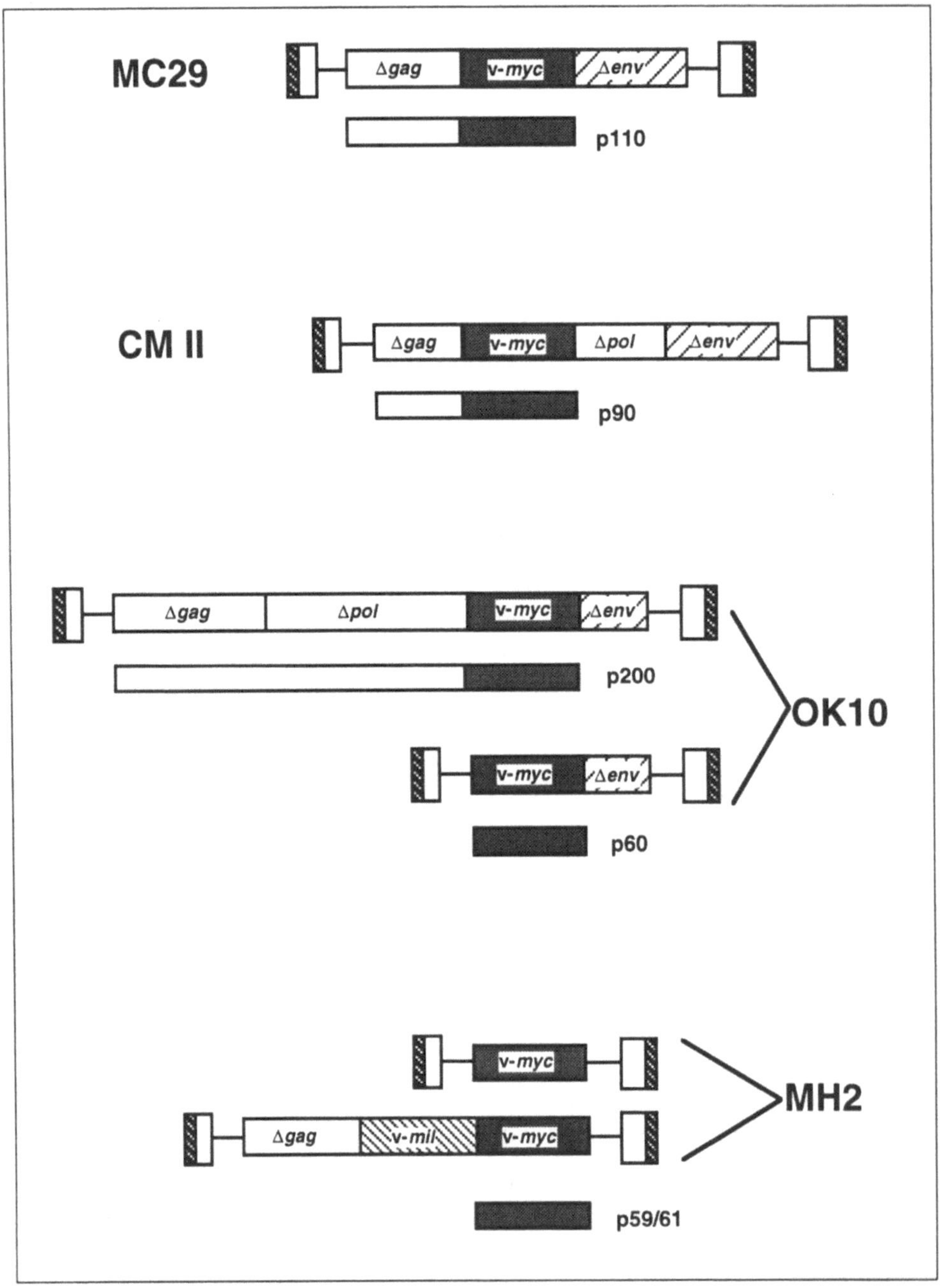

Fig. 3.1. Alleles of v-myc *and their protein product. Genomic structures of* v-myc *from strains MC29, CM II, OK10 and MH2 are shown with the corresponding protein products below each genome. Note that* v-mil *produces a v-Raf protein.*

sequence replaces all of *pol* and the 3' and 5' sequences of *gag* and *env*, respectively. Likewise, the genetic structures of CMII, OK10 and MH2 were determined to contain alleles of v-*myc*.

The v-*myc* allele of MC29 produces a polypeptide of 110 kDa that consists of gag protein sequence fused to the N-terminal domain of Myc.[8-14] Likewise, the CMII allele produces a p90 gag-Myc fusion protein.[12,15,16] Expression of the OK10 *myc* allele results in a gag-pol-Myc p200 fusion protein as well as a p60 protein consisting of mostly Myc sequences.[17-20] The MH2 strain also carries another oncogene termed v-*mil* (later found to be identical to *raf*) that is located 5' of v-*myc* and encodes a p59/61 v-Myc protein.[15,19,21,22] The FH3 strain was isolated after injection of a 10-day-old chick embryo with avian leukosis virus.[6] The v-*myc* gene of FH3 is expressed as a p145 gag-Myc protein.[23] The functions of some of these proteins will be discussed in subsequent chapters.

A new paradigm for cancer genetics emerged when the existence of normal chicken cellular DNA sequences related to v-*src* and subsequently to v-*myc* oncogene of MC29 were identified by nucleic hybridization experiments (Fig. 3.2).[24-26] The chicken c-*myc* gene was subsequently molecularly cloned and sequenced.[27] It contains three exons, the second and third of which encodes a protein predicted to be 416 amino acids in length and a predicted

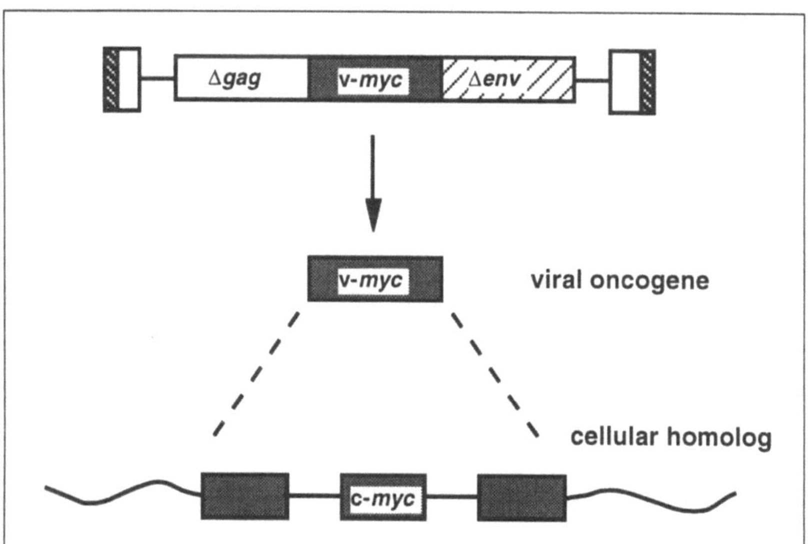

Fig. 3.2. Identification of c-myc *using retroviral* myc *as a probe.*

M_r of 46,000. The cDNA and genomic DNA clones of human c-*myc* were subsequently isolated and found to share significant homology with the chicken gene.[28-36]

3.3 INSERTIONAL ACTIVATION OF c-*myc* BY AVIAN LEUKOSIS VIRUS

The discovery of the normal cellular c-*myc* gene raises the question of whether the proto-oncogene could be activated in ways other than being captured by a retrovirus. Indeed, in the majority of bursal (B-cell) lymphomas in chickens induced by avian leukosis viral infection, the provirus integrated around or within the c-*myc* locus.[37,38] These tumors displayed a high level of expression of c-*myc* mRNA. In many cases, proviral integration occurred in intron 2 in the same orientation as the *myc* transcription unit. These observations suggested that promoter insertional activation of c-*myc* results in elevated levels of *myc* mRNA and contributes to bursal lymphomagenesis.[39,40] Thus, the human counterpart to retroviral insertional activation of c-*myc* in avian lymphoma may arise from activation by chromosomal translocation that juxtaposes strong positive transcriptional regulatory elements near the *myc* locus.

REFERENCES

1. Bishop JM. Retroviruses and cancer genes. [Review]. Adv Cancer Res 1982; 37:1-32.
2. Bishop JM. Cellular oncogenes and retroviruses. [Review]. Annu Rev Biochem 1983; 52:301-54.
3. Bishop JM. Retroviruses. [Review]. Annu Rev Biochem 1978; 47:35-88.
4. Bishop JM. Nobel Lecture. Retroviruses and oncogenes II. [Review]. Biosci Rep 1990; 10:473-91.
5. Bister K, Jansen HW. Oncogenes in retroviruses and cells: biochemistry and molecular genetics. [Review]. Adv Cancer Res 1986; 47:99-188.
6. Chen C, Biegalke BJ, Eisenman RN, Linial ML. FH3, a v-myc avian retrovirus with limited transforming ability. J Virol 1989; 63:5092-100.
7. Tikhonenko AT, Linial ML. gag as well as myc sequences contribute to the transforming phenotype of the avian retrovirus FH3. J Virol 1992; 66:946-55.
8. Sheiness D, Fanshier L, Bishop JM. Identification of nucleotide sequences which may encode the oncogenic capacity of avian retrovirus MC29. J Virol 1978; 28:600-10.

9. Moelling K, Owada MK, Greiser-Wilke I, Bunte T, Donner P. Biochemical characterization of transformation-specific proteins of acute avian leukemia and sarcoma viruses. J Cell Biochem 1982; 20:63-9.
10. Sovova V, Travnicek M, Hlozanek I, Cerna H. In vitro cleavage of gag-myc fused protein P110 of a defective leukaemia virus MC29 by retroviral protease p15. Folia Biol 1982; 28:369-76.
11. Reddy EP, Reynolds RK, Watson DK, Schultz RA, Lautenberger J, Papas TS. Nucleotide sequence analysis of the proviral genome of avian myelocytomatosis virus (MC29). Proc Natl Acad Sci USA 1983; 80:2500-4.
12. Bunte T, Greiser-Wilke I, Moelling K. The transforming protein of the MC29-related virus CMII is a nuclear DNA-binding protein whereas MH2 codes for a cytoplasmic RNA-DNA binding polyprotein. EMBO J 1983; 2:1087-92.
13. Alitalo K, Bishop JM, Smith DH, Chen EY, Colby WW, Levinson AD. Nucleotide sequence to the v-myc oncogene of avian retrovirus MC29. Proc Natl Acad Sci USA 1983; 80:100-4.
14. Bister K, Trachmann C, Jansen HW, Schroeer B, Patschinsky T. Structure of mutant and wild-type MC29 v-myc alleles and biochemical properties of their protein products. Oncogene 1987; 1:97-109.
15. Saule S, Coll J, Righi M, Lagrou C, Raes MB, Stehelin D. Two different types of transcription for the myelocytomatosis viruses MH2 and CMII. EMBO J 1983; 2:805-9.
16. Walther N, Lurz R, Patschinsky T, Jansen HW, Bister K. Molecular cloning of proviral DNA and structural analysis of the transduced myc oncogene of avian oncovirus CMII. J Virol 1985; 54:576-85.
17. Chiswell DJ, Ramsay G, Hayman MJ. Two virus-specific rna species are present in cells transformed by defective leukemia virus OK10. J Virol 1981; 40:301-4.
18. Hann SR, Abrams HD, Rohrschneider LR, Eisenman RN. Proteins encoded by v-myc and c-myc oncogenes: identification and localization in acute leukemia virus transformants and bursal lymphoma cell lines. Cell 1983; 34:789-98.
19. Patschinsky T, Walter G, Bister K. Immunological analysis of v-myc gene products using antibodies against a myc-specific synthetic peptide. Virology 1984; 136:348-58.
20. Bunte T, Donner P, Pfaff E, et al. Inhibition of DNA binding of purified p55v-myc in vitro by antibodies against bacterially expressed myc protein and a synthetic peptide. EMBO J 1984; 3:1919-24.
21. Jansen HW, Patschinsky T, Bister K. Avian oncovirus MH2: molecular cloning of proviral DNA and structural analysis of viral RNA and protein. J Virol 1983; 48:61-73.

22. Kan NC, Flordellis CS, Garon CF, Duesberg PH, Papas TS. Avian carcinoma virus MH2 contains a transformation-specific sequence, mht, and shares the myc sequence with MC29, CMII, and OK10 viruses. Proc Natl Acad Sci USA 1983; 80:6566-70.
23. Tikhonenko AT, Linial ML. Transforming variants of the avian myc-containing retrovirus FH3 arise prior to phenotypic selection. J Virol 1993; 67:3635-8.
24. Parker RC, Varmus HE, Bishop JM. Cellular homologue (c-src) of the transforming gene of Rous sarcoma virus: isolation, mapping, and transcriptional analysis of c-src and flanking regions. Proc Natl Acad Sci USA 1981; 78:5842-6.
25. Sheiness DK, Hughes SH, Varmus HE, Stubblefield E, Bishop JM. The vertebrate homolog of the putative transforming gene of avian myelocytomatosis virus: characteristics of the DNA locus and its RNA transcript. Virology 1980; 105:415-24.
26. Lautenberger JA, Schulz RA, Garon CF, Tsichlis PN, Papas TS. Molecular cloning of avian myelocytomatosis virus (MC29) transforming sequences. Proc Natl Acad Sci USA 1981; 78:1518-22.
27. Vennstrom B, Sheiness D, Zabielski J, Bishop JM. Isolation and characterization of c-myc, a cellular homolog of the oncogene (v-myc) of avian myelocytomatosis virus strain 29. J Virol 1982; 42:773-9.
28. Collins S, Groudine M. Amplification of endogenous myc-related DNA sequences in a human myeloid leukaemia cell line. Nature 1982; 298:679-81.
29. Dalla-Favera R, Gelmann EP, Martinotti S, et al. Cloning and characterization of different human sequences related to the onc gene (v-myc) of avian myelocytomatosis virus (MC29). Proc Natl Acad Sci USA 1982; 79:6497-501.
30. Dalla-Favera R, Bregni M, Erikson J, Patterson D, Gallo RC, Croce CM. Human c-myc onc gene is located on the region of chromosome 8 that is translocated in Burkitt lymphoma cells. Proc Natl Acad Sci USA 1982; 79:7824-7.
31. Watson DK, Psallidopoulos MC, Samuel KP, Dalla-Favera R, Papas TS. Nucleotide sequence analysis of human c-myc locus, chicken homologue, and myelocytomatosis virus MC29 transforming gene reveals a highly conserved gene product. Proc Natl Acad Sci USA 1983; 80:3642-5.
32. Dalla-Favera R, Martinotti S, Gallo RC, Erikson J, Croce CM. Translocation and rearrangements of the c-myc oncogene locus in human undifferentiated B-cell lymphomas. Science 1983; 219:963-7.
33. Nowell P, Finan J, Dalla-Favera R, et al. Association of amplified oncogene c-myc with an abnormally banded chromosome 8 in a human leukaemia cell line. Nature 1983; 306:494-7.
34. Gelmann EP, Psallidopoulos MC, Papas TS, Dalla-Favera R. Identification of reciprocal translocation sites within the c-myc oncogene and immunoglobulin mu locus in a Burkitt lymphoma. Nature 1983; 306:799-803.

35. Battey J, Moulding C, Taub R, et al. The human c-myc oncogene: structural consequences of translocation into the IgH locus in Burkitt lymphoma. Cell 1983; 34:779-87.
36. Watt R, Stanton LW, Marcu KB, Gallo RC, Croce CM, Rovera G. Nucleotide sequence of cloned cDNA of human c-myc oncogene. Nature 1983; 303:725-8.
37. Hayward WS, Neel BG, Astrin SM. Activation of a cellular onc gene by promoter insertion in ALV-induced lymphoid leukosis. Nature 1981; 290:475-80.
38. Neel BG, Hayward WS, Robinson HL, Fang J, Astrin SM. Avian leukosis virus-induced tumors have common proviral integration sites and synthesize discrete new RNAs: oncogenesis by promoter insertion. Cell 1981; 23:323-34.
39. Payne GS, Bishop JM, Varmus HE. Multiple arrangements of viral DNA and an activated host oncogene in bursal lymphomas. Nature 1982; 295:209-14.
40. Pachl C, Schubach W, Eisenman R, Linial M. Expression of c-myc RNA in bursal lymphoma cell lines: identification of c-myc-encoded proteins by hybrid-selected translation. Cell 1983; 33:335-44.

CHAPTER 4

Structure of the c-*myc* Gene and its Transcription

4.1 c-*myc* GENE AND mRNA STRUCTURES

The c-*myc* gene belongs to a family of related genes that encode for the Myc proteins: c-Myc, L-Myc, N-Myc and s-Myc. L-*myc* was identified as a highly expressed gene in human lung cancer.[1] N-*myc* gene is found with amplified copy number in human neuroblastoma.[2] A rat gene, s-*myc*, was cloned via homology with c-*myc*.[3] Intriguingly, studies of s-*myc* suggest that it suppresses neoplastic transformation.[4,5] All Myc family members share conserved sequences in the N-terminal region of the protein. At least two conserved boxes have been recognized and termed Myc box I (amino acids 55-65 of c-Myc) and Myc box II (amino acids 122-144 of c-Myc).

The general genomic organization and coding sequences of c-*myc* have been highly conserved throughout vertebrate evolution from zebrafish to man.[6,7] The coding sequence is conserved down to the invertebrate seastar.[8] Normal c-*myc* transcription units are composed of three exons with the second and third encoding the major Myc protein species (Fig. 4.1).[9] These two exons have at least 70% sequence identity between species. All c-*myc* genes have a long untranslated exon 1 whose sequence is not as well conserved through evolution. The first exon has been postulated to play a role in translational control and mRNA stability; it also contains the site for attenuation of transcriptional elongation.[10,11]

There are multiple promoters for the c-*myc* transcription unit.[11] The two major promoters, P1 and P2, are located about 160 bp apart within exon 1 (Fig. 4.1). In normal cells the P2 promoter accounts for 75% of the transcripts, and P1 accounts for the majority of the remaining 25%. Another promoter, P0, located about

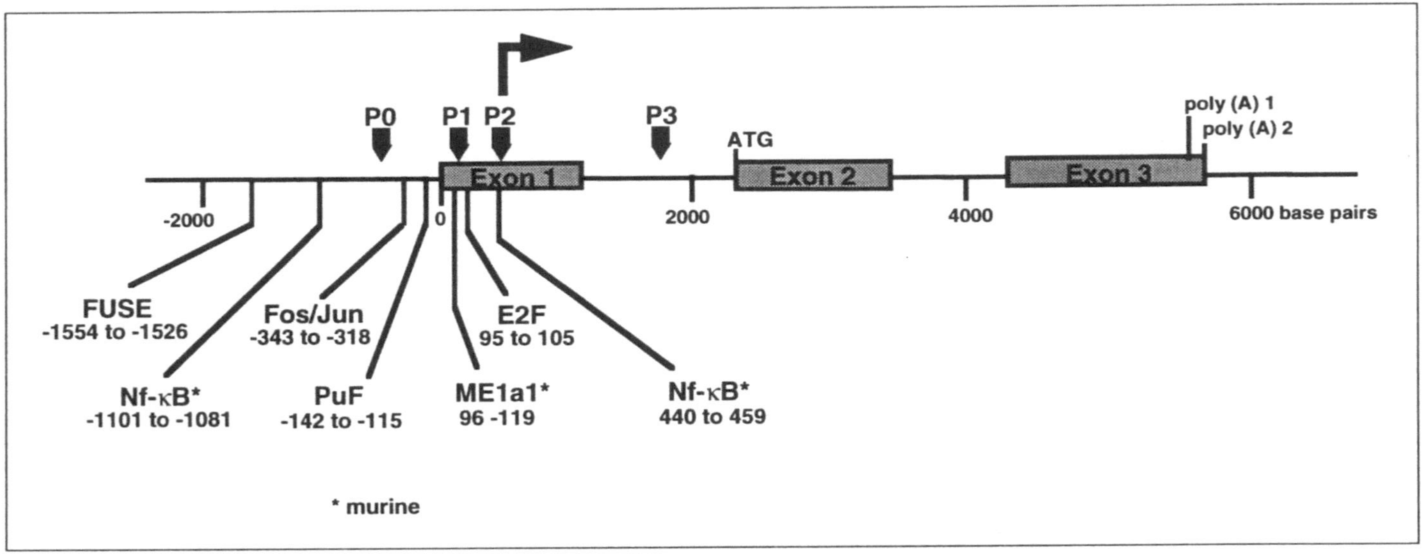

Fig. 4.1. Genomic organization of c-myc. c-myc is shown with its three exons spanning about 4 Kb. The promoters P0, P1, P2 (which is most frequently used) and P3 are indicated. Sites of selected cis elements that are known to modulate c-myc transcription are shown. Two polyadenylation signals are shown in exon 3.

600 bp upstream from P1 accounts for about 5% of the mRNAs. Several additional minor transcripts originating from noncanonical TATA sequences have been detected. Another promoter, P3, has been mapped to the 3' end of intron 1. Cryptic start sites near P3 may account for transcription initiation from translocated c-*myc* alleles such as those found in Burkitt lymphoma. Two conserved polyadenylation sites have been identified and located about 150 bp apart 3' of exon 3.

Mature c-*myc* mRNAs initiated from P2 and P1 are 2.2 and 2.4 kb, respectively.[10,11] Those initiated from P0 and P3 are 3.1 and 2.3 kb, respectively. Intriguingly, antisense transcripts originating from several regions of the gene have been detected; however, the functional role of these transcripts are unknown.

4.2 REGULATION OF *c-myc* TRANSCRIPTION INITIATION

Regulation of c-*myc* transcription initiation is complex since there are multiple promoters.[10,11] A number of positive and negative *cis* acting elements have been identified and shown to be important for transcription from P1 and P2 promoters at which 90% of the transcripts are initiated. A number of different approaches have been employed to dissect the promoter regions of c-*myc* including microinjection of DNA constructs into *Xenopus* oocytes or transfection of plasmid constructs into mammalian cells. Well over a dozen positive and negative *cis* regulatory elements have been identified (Fig. 4.1), some of which are particularly intriguing with respect to cancer biology.

Two positive *cis* elements, the nuclease hypersensitive element (NHE) and far upstream element (FUSE), are regions of c-*myc* that are nuclease hypersensitive and have potential to form single strand or triplex DNA. Efficient transcription of c-*myc* from P1 and P2 requires sequences between -142 and -115 (from P1) that contain a putative triple helical (H-DNA) purine/pyrimidine sequence or NHE. This sequence is bound by the transcription factor PuF (*nm23*-H2) which was recently shown to be homologous to the product of *nm23* (*nm23*-H1), a gene that has been implicated in suppression of cancer metastasis.[12] Although *nm23* has been found to be a dinucleoside phosphate kinase and homologous to the *Drosophila* gene, *awd,* which is involved in wing development, its homology to a potential transcription factor is indeed surprising and intriguing. The nucleoside diphosphate kinase activity is not required for its transcriptional activity.[13] These findings suggest that c-*myc* activation by *nm23*-H2 may be a critical event in cancer metastasis; however, this is purely speculative at this point.

The far upstream element (FUSE) was identified through the use of a sensitive exonuclease assay that detects, in vitro, binding of sequence specific factors with c-*myc* upstream elements.[14] Binding at one site, more than 1500 bp upstream of the c-*myc* promoter P1, is abrogated with dimethyl sulfoxide (DMSO) induced differentiation of HL-60 and U-937 leukemia cells. Since the binding activity was highest before the decrease of c-*myc* expression during differentiation, it was surmised that the FUSE potentiates c-*myc* expression. A 4-base pair deletion that eliminated binding to FUSE reduced expression of a transfected, chimeric c-*myc*-CAT gene in untreated, but not in DMSO-treated U-937 cells.

The FUSE-binding protein (FBP), which is present in undifferentiated but not differentiated cells, was purified.[15] Based on peptide sequences from the purified protein, a cDNA encoding FBP was cloned. FBP mRNA level declined upon differentiation, suggesting that FBP is regulated transcriptionally. Recombinant FBP forms sequence-specific complexes with a single strand of FUSE, and when transfected into human leukemia cells, FBP stimulated a c-*myc*-promoter-reporter plasmid in a FUSE-dependent manner. The FBP structure defines a novel single-strand DNA-binding protein, which is able to force strand separation of short stretches of linear double-stranded (dsDNA).[16] The link between regional dsDNA melting with *cis*-element binding by a transcriptional activator suggests a novel mechanism for the regulation of eukaryotic gene expression.

A site within murine c-*myc* exon 1, located at +440 to +459 bp relative to the P1 promoter, contains an NF-κB-like binding element.[17] This site forms a variety of complexes with nuclear proteins derived from the murine WEHI 231 B-lymphoma cells in exponential growth. These complexes are altered following treatment with phorbol ester, when transcription of c-*myc* is reduced. Mutation of this site within the context of the c-*myc* promoter prevented NF-κB-mediated induction of transcription of this oncogene. Subsequently, a second site was identified to which *rel*/NFκB family members could bind.[18] Factor binding to these two

NF-κB elements mediates induction of transcription of the c-*myc* promoter upon interleukin-l treatment of human dermal fibroblasts and human T cell leukemia virus type I *tax* gene expression in T cells.

In the human c-*myc* gene, an in vivo footprint over potential NF-κB sites in the 5' flanking sequence of c-*myc* and in the first exon of the c-*myc* gene has been identified on the translocated alleles in the Burkitt lymphoma cell lines, Ramos and Raji, respectively.[19] In electrophoretic mobility shift assays, a supershift was observed with anti-p50 (NF-κB) antibody with the exon 1 site. The 5'-site shift complex disappeared with anti-p50 antibody. These sites are blocked in the silent normal c-*myc* allele in Burkitt lymphoma cells, whereas Rel family proteins bind to these sites in the translocated *myc* allele. Studies of both human and murine c-*myc* suggest that two NF-κB sites function as positive regulatory regions for the c-*myc* gene.

About 60 bp 5' to the promoter P2 is the binding site GGCGGGAAA for the transcription factor E2F that has now been shown to interact with the retinoblastoma protein, pRb.[20-24] The E2F binding site mediates the transactivation of c-*myc* by the adenovirus E1A protein. The interaction of pRb with E2F abrogates the ability of E2F to activate transcription. Thus, among the many functions of pRb may be its ability to downregulate the expression of c-*myc*. Studies of TGFβ inhibition of c-*myc* transcription suggest that pRb might be involved in suppressing E2F in this pathway.[23,25,26] However, TGFβ appears to be able to downregulate c-*myc* in pRb negative cell lines as well. It has not been ruled out that the pRb-like p107 protein might mediate inhibition by TGFβ in pRb negative cell lines.

Between -343 and -318bp from P1 is a 26 bp negative regulatory element bearing binding sites for *fos/jun* and octamer binding protein 1 (Oct-1).[27] This sequence has been shown to bind octamer binding protein in vitro and a binding activity from HeLa nuclear extract was able to be competed by a consensus AP-1 site.[28,29] While several experimental approaches suggest that this sequence behaves as a negative regulatory element, the physiological significance of *fos/jun* mediated inhibition of c-*myc* transcription is unclear.

The human Cut protein has been identified as a negative regulator of c-*myc* transcription through the ME1a1 site that is located upstream of the P2 TATA box.[30] A cDNA encoding a protein that binds the ME1a1 site was isolated by screening of an expression library with an ME1a1 DNA probe. This cDNA encodes the human homolog of the *Drosophila* Cut homeodomain protein. Antibodies against human Cut abrogates ME1a1 binding activity in nuclear extracts. In cotransfection experiments, the human Cut protein repressed transcription from the c-*myc* promoter in an ME1a1 site dependent manner. c-*myc* exon 1 sequences were also necessary, in addition to the ME1a1 site, for repression by Cut. A zinc finger protein, termed MAZ, has also been found to bind to the ME1a1 site.[31] Although it has been implicated in c-*myc* transcriptional attenuation, evidence to support this hypothesis is still pending.

The differential utilization of P1 versus P2 promoters (frequency of 1 to 4) has been suggested to be influenced by promoter elements found in between and around these promoters. Specifically, there is one CT-box (CCCTCCCC) between P1 and P2, and five repeats of CT-box are found between -151 and -101 from P1.[32] All CT-boxes are required for efficient transcription from P2. Recent mutagenesis studies suggest that occupancy of the CT site between P1 and P2 by Sp1 or Sp1-like factors causes preferential usage of P2 over P1. Presumably, this results from a block by the Sp1-like factor that prevents elongation of transcripts initiated from P1. With a strong heterologous enhancer placed 5' of P1 and limiting amounts of CT-box binding factors, transcription is preferentially initiated from P1. Indeed, this is similar to the pathologic situation with Burkitt lymphomas in which strong immunoglobulin enhancers are juxtaposed to the c-*myc* transcription unit associated with the preferential usage of P1. The functional consequence of P1 promoter usage is the loss of transcriptional elongation attenuation (discussed below) resulting in elevated c-*myc* mRNAs.

4.3 REGULATION OF c-*myc*: TRANSCRIPTIONAL ELONGATION

Regulation of transcriptional elongation by decreasing the rate of extension of already initiated mRNAs (transcriptional attenuation) is a well-known mechanism in prokaryotes that has gained much attention as a mechanism for the control of eukaryotic transcription through studies of c-*myc*.[10,11,33] Early studies of the promyelocytic leukemia cell line, HL-60, indicated that retinoic acid-induced differentiation of these cells to mature granulocytes is associated with a marked decrease in c-*myc* mRNA. Most of the decrease, however, was not due to alterations in c-*myc* transcription initiation rates but due to an increase in transcriptional attenuation localized at the end of exon 1 (Fig. 4.1). The increase in c-*myc* transcripts with mitogenic stimulation of mononuclear cells is due to a release of the block to transcription elongation at the 3' end of exon 1. The structural element at the 3' end of exon 1 appears to be a putative stem-loop structure that may be homologous to the prokaryotic *rho*-independent transcriptional terminating DNA sequences. The means by which preferential usage of the P1 promoter alleviates transcriptional attenuation remain unknown. Nevertheless, these altered mechanisms of regulation of translocated c-*myc* alleles in Burkitt's lymphoma may be of fundamental pathophysiological importance.

A loss of the block to transcription elongation in exon 1 and a promoter shift to use of the upstream P1 promoter are hallmarks of many translocated c-*myc* alleles in Burkitt's lymphoma.[34] The mutations within and surrounding several Burkitt's lymphoma c-*myc* alleles are not sufficient to abrogate the transcription elongation block. However, transcription initiation from the P2 promoter may be necessary to obtain the block to transcription elongation. Deletion mutagenesis of the P1 or P2 promoters was used to determine if specific promoter usage is necessary for the block in transcription elongation. P1-initiated c-*myc* transcripts do not terminate at discrete sites near the 3' end of exon 1, whereas P2-initiated transcripts either terminate or read through the transcription block signals. Overexpression or constitutive expression from the c-*myc* P1 promoter, therefore, may contribute to increased transcriptional readthrough in Burkitt's lymphoma cells. In contrast, the ability of normal cells to modulate P2-initiated transcription through blocking elongation provides a fine control over c-*myc* RNA levels.

Although great strides have been achieved in identifying negative and positive *cis* elements and their binding proteins that affect c-*myc* transcription, it still remains unclear how these various factors regulate normal c-*myc* expression, for example after serum stimulation. The state-of-the-art in the study of c-*myc* transcriptional regulation is like the recruitment of talented symphony musicians who await the right conductor to orchestrate their activities. Thus, a synthesis of these discoveries in the context of biological phenomena appears to be necessary for full appreciation of the transcription factors that regulate c-*myc* expression. Nevertheless, these studies have yielded an initial picture of how aberrant c-*myc* transcription occurs in Burkitt's lymphoma. A challenge for future studies is the identification of signal transduction pathways and the *cis* elements that directly connect c-*myc* expression to extracellular signals, whether they be growth factors or components of the extracellular matrix.

4.4 REGULATION of c-*myc* mRNA STABILITY

In addition to the complex nature of transcription initiation and elongation of c-*myc*, modulation of the half-life of the c-*myc* mRNA provides another mechanism by which steady-state c-*myc* mRNA is regulated.[11] The c-*myc* mRNA half-life is known to be modulated by serum deprivation, cell contact inhibition, differentiation induction, or mitogenic stimulation.[35-38] For example, transcriptional attenuation causes the initial decrease in c-*myc* mRNA in murine erythroleukemia cells induced to differentiate, but rapid degradation decreases c-*myc* mRNA level at later stages. In contrast, the increase in c-*myc* mRNA accompanying renal regeneration is largely due to enhanced mRNA stability.

The half-life of c-*myc* mRNA in normal cycling cells varies from 10 minutes to less than 1 hour.[39] The half-lives of transcripts initiated from the P1 or P2 promoters are similar. The deadenylated P1 and P2 transcripts are both more stable than the corresponding adenylated mRNAs.[40,41] This is a clear exception to the general notion that polyadenylation contributes to mRNA stability. The 3' untranslated sequences of c-*myc* have been implicated in its rapid turnover.[42,43] Several studies have uncovered sequences, such as the 3' untranslated U-rich sequence, within the c-*myc* mRNA that contribute to its instability.[42-44] Other findings, however, suggest that the U-rich sequence is dispensable for rapid degradation of c-*myc* mRNA.[45] Genetic alterations that remove the 3' sequences of c-*myc* resulted in a prolonged mRNA half-life.[46] Recently, Wisdom and Lee demonstrated that the c-*myc* coding sequence localized to exon 3 that encodes the c-Myc protein DNA-binding domain is a major determinant of mRNA instability.[47] Intriguingly, evidence was also presented suggesting that translation is required for this sequence in exon 3 to confer mRNA instability.[47,48] This phenomenon is reminiscent of the requirement of translation for tubulin mRNA degradation.[49,50]

REFERENCES

1. Nau MM, Brooks BJ, Battey J, et al. L-myc, a new myc-related gene amplified and expressed in human small cell lung cancer. Nature 1985; 318:69-73.
2. Schwab M, Alitalo K, Klempnauer KH, et al. Amplified DNA with limited homology to myc cellular oncogene is shared by human neuroblastoma cell lines and a neuroblastoma tumour. Nature 1983; 305:245-8.
3. Sugiyama A, Kume A, Nemoto K, et al. Isolation and characterization of s-myc, a member of the rat myc gene family. Proc Natl Acad Sci U S A 1989; 86:9144-8.
4. Lee SY, Sugiyama A, Sueoka N, Kuchino Y. Point mutation of the neu gene in rat neural tumor RT4-AC cells: suppression of tumorigenicity by s-Myc. Jpn J Cancer Res 1990; 81:1085-8.
5. Asai A, Miyagi Y, Sugiyama A, et al. The s-Myc protein having the ability to induce apoptosis is selectively expressed in rat embryo chondrocytes. Oncogene 1994; 9:2345-52.
6. Kato GJ, Dang CV. Function of the c-Myc oncoprotein. [Review]. FASEB J 1992; 6:3065-72.
7. Meichle A, Philipp A, Eilers M. The functions of Myc proteins. [Review]. Biochim Biophys Acta 1992; 1114:129-46.
8. Walker CW, Boom JD, Marsh AG. First non-vertebrate member of the myc gene family is seasonally expressed in an invertebrate testis. Oncogene 1992; 7:2007-12.
9. Battey J, Moulding C, Taub R, et al. The human c-myc oncogene: structural consequences of translocation into the IgH locus in Burkitt lymphoma. Cell 1983; 34:779-87.
10. Spencer CA, Groudine M. Control of c-myc regulation in normal and neoplastic cells. [Review]. Adv Cancer Res 1991; 56:1-48.
11. Marcu K, Bossone S, Patel A. Myc function and regulation. Ann Rev Biochem 1992; 61:809-860.
12. Postel EH, Berberich SJ, Flint SJ, Ferrone CA. Human c-myc transcription factor PuF identified as nm23-H2 nucleoside diphosphate kinase, a candidate suppressor of tumor metastasis. Science 1993; 261:478-80.
13. Postel EH, Ferrone CA. Nucleoside diphosphate kinase enzyme activity of NM23-H2/PuF is not required for its DNA binding and in vitro transcriptional functions. J Biol Chem 1994; 269:8627-30.
14. Avigan MI, Strober B, Levens D. A far upstream element stimulates c-myc expression in undifferentiated leukemia cells. J Biol Chem 1990; 265:18538-45.

15. Duncan R, Bazar L, Michelotti G, et al. A sequence-specific, single-strand binding protein activates the far upstream element of c-myc and defines a new DNA-binding motif. Genes Dev 1994; 8:465-80.
16. Bazar L, Meighen D, Harris V, Duncan R, Levens D, Avigan M. Targeted melting and binding of a DNA regulatory element by a transactivator of c-myc. J Biol Chem 1995; 270:8241-8.
17. Kessler DJ, Spicer DB, La Rosa FA, Sonenshein GE. A novel NF-kappa B element within exon 1 of the murine c-myc gene. Oncogene 1992; 7:2447-53.
18. Larosa FA, Pierce JW, Sonenshein GE. Differential regulation of the c-myc oncogene promoter by the NFκB rel family of transcription factors. Mol Cell Biol1994; 14:1039-44.
19. Ji L, Arcinas M, Boxer LM. NF-kappa B sites function as positive regulators of expression of the translocated c-myc allele in Burkitt's lymphoma. Mol Cell Biol 1994; 14:7967-74.
20. Oswald F, Lovec H, Moroy T, Lipp M. E2F-dependent regulation of human MYC: trans-activation by cyclins D1 and A overrides tumour suppressor protein functions. Oncogene 1994; 9:2029-36.
21. Ishida S, Shudo K, Takada S, Koike K. Transcription from the P2 promoter of human protooncogene myc is suppressed by retinoic acid through an interaction between the E2F element and its binding proteins. Cell Growth Differ 1994; 5:287-94.
22. Roussel MF, Davis JN, Cleveland JL, Ghysdael J, Hiebert SW. Dual control of myc expression through a single DNA binding site targeted by ets family proteins and E2F-1. Oncogene 1994; 9:405-15.
23. Satterwhite DJ, Aakre ME, Gorska AE, Moses HL. Inhibition of cell growth by TGF-beta-1 is associated with inhibition of B-myb and cyclin A in both BALB/MK and MV1LU cells. Cell Growth Diff 1994; 5:789-99.
24. Stasiv YZ, Mashkova TD, Chernov BK, Sokolova IV, Itkes AV, Kisselev LL. Cloning of a cDNA encoding a human protein which binds a sequence in the c-myc gene similar to the interferon-stimulated response element. Gene 1994; 145:267-72.
25. Pietenpol JA, Stein RW, Moran E, et al. TGF-beta 1 inhibition of c-myc transcription and growth in keratinocytes is abrogated by viral transforming proteins with pRB binding domains. Cell 1990; 61:777-85.
26. Robinson-Benion C, Salhany KE, Hann SR, Holt JT. Antisense inhibition of c-myc expression reveals common and distinct mechanisms of growth inhibition by TGF beta and TNF alpha. J Cell Biochem 1991; 45:188-95.
27. Hay N, Bishop JM, Levens D. Regulatory elements that modulate expression of human c-myc. Genes Dev 1987; 1:659-71.
28. Takimoto M, Quinn JP, Farina AR, Staudt LM, Levens D. fos/jun and octamer-binding protein interact with a common site in a negative element of the human c-myc gene. J Biol Chem 1989; 264:8992-9.

29. Hay N, Takimoto M, Bishop JM. A FOS protein is present in a complex that binds a negative regulator of MYC. Genes Dev 1989; 3:293-303.
30. Dufort D, Nepveu A. The human cut homeodomain protein represses transcription from the c-myc promoter. Mol Cell Biol 1994; 14:4251-4257.
31. Bossone SA, Asselin C, Patel AJ, Marcu KB. MAZ, a zinc finger protein, binds to c-MYC and C2 gene sequences regulating transcriptional initiation and termination. Proc Natl Acad Sci U S A 1992; 89:7452-6.
32. DesJardins E, Hay N. Repeated CT elements bound by zinc finger proteins control the absolute and relative activities of the two principal human c-myc promoters. Mol Cell Biol 1993; 13:5710-24.
33. Spencer CA, Groudine M. Transcription elongation and eukaryotic gene regulation. [Review]. Oncogene 1990; 5:777-85.
34. Spencer CA, LeStrange RC, Novak U, Hayward WS, Groudine M. The block to transcription elongation is promoter dependent in normal and Burkitt's lymphoma c-myc alleles. Genes Dev 1990; 4:75-88.
35. Blanchard JM, Piechaczyk M, Dani C, et al. c-myc gene is transcribed at high rate in G0-arrested fibroblasts and is post-transcriptionally regulated in response to growth factors. Nature 1985; 317:443-5.
36. Dani C, Mechti N, Piechaczyk M, Lebleu B, Jeanteur P, Blanchard JM. Increased rate of degradation of c-myc mRNA in interferon-treated Daudi cells. Proc Natl Acad Sci U S A 1985; 82:4896-9.
37. Jonak GJ, Knight E, Jr. Selective reduction of c-myc mRNA in Daudi cells by human beta interferon. Proc Natl Acad Sci U S A 1984; 81:1747-50.
38. Piechaczyk M, Yang JQ, Blanchard JM, Jeanteur P, Marcu KB. Posttranscriptional mechanisms are responsible for accumulation of truncated c-myc RNAs in murine plasma cell tumors. Cell 1985; 42:589-97.
39. Dani C, Blanchard JM, Piechaczyk M, El Sabouty S, Marty L, Jeanteur P. Extreme instability of myc mRNA in normal and transformed human cells. Proc Natl Acad Sci U S A 1984; 81:7046-50.
40. Swartwout SG, Preisler H, Guan WD, Kinniburgh AJ. Relatively stable population of c-myc RNA that lacks long poly(A). Mol Cell Biol 1987; 7:2052-8.
41. Swartwout SG, Kinniburgh AJ. c-myc RNA degradation in growing and differentiating cells: possible alternate pathways. Mol Cell Biol 1989; 9:288-95.
42. Jones TR, Cole MD. Rapid cytoplasmic turnover of c-myc mRNA: requirement of the 3' untranslated sequences. Mol Cell Biol 1987; 7:4513-21.

43. Bonnieu A, Piechaczyk M, Marty L, et al. Sequence determinants of c-myc mRNA turn-over: influence of 3' and 5' non-coding regions. Oncogene Res 1988; 3:155-66.
44. Shaw G, Kamen R. A conserved AU sequence from the 3' untranslated region of GM-CSF mRNA mediates selective mRNA degradation. Cell 1986; 46:659-67.
45. Bonnieu A, Roux P, Marty L, Jeanteur P, Piechaczyk M. AUUUA motifs are dispensable for rapid degradation of the mouse c-myc RNA. Oncogene 1990; 5:1585-8.
46. Aghib DF, Bishop JM, Ottolenghi S, Guerrasio A, Serra A, Saglio G. A 3' truncation of MYC caused by chromosomal translocation in a human T-cell leukemia increases mRNA stability. Oncogene 1990; 5:707-11.
47. Wisdom R, Lee W. The protein-coding region of c-myc mRNA contains a sequence that specifies rapid mRNA turnover and induction by protein synthesis inhibitors. Genes Dev 1991; 5:232-43.
48. Wisdom R, Lee W. Translation of c-myc mRNA is required for its post-transcriptional regulation during myogenesis. J Biol Chem 1990; 265:19015-21.
49. Pachter JS, Yen TJ, Cleveland DW. Autoregulation of tubulin expression is achieved through specific degradation of polysomal tubulin mRNAs. Cell 1987; 51:283-92.
50. Gay DA, Yen TJ, Lau JT, Cleveland DW. Sequences that confer beta-tubulin autoregulation through modulated mRNA stability reside within exon 1 of a beta-tubulin mRNA. Cell 1987; 50:671-9.

CHAPTER 5

INVOLVEMENT OF c-*myc* IN HUMAN CANCERS

5.1 INTRODUCTION

After 400 million years of evolutionary selection pressures and spontaneous order through the force of self-organization, the genome of organisms is safeguarded by complex systems of regulatory molecules from lethal environmental damages. Only over the past century, the by-products of civilization and luxury have infiltrated these safeguards. Certain environmental exposures, such as radiation and smoking among others, have clearly been implicated for their ability to induce human cancers. Cancer has also been shown to have a hereditary component as well, since first recognized in the family histories of cancer patients and now confirmed by the identification of germline neoplastic mutations. Virtually all conceivable alterations of DNA have been detected in human cancers. These include gene amplifications, deletions, point mutations, or translocations that affect gene expression. Recently, alterations in the DNA damage repair machinery are recognized as hallmarks of human cancers.[1] It is currently thought that genetic alterations involved in human cancers result in several possible biological outcomes that contribute to the malignant phenotype. Since the normal signal transduction pathways may influence cell proliferation positively through expression of proto-oncogenes or negatively through expression of tumor suppressor genes, loss of tumor suppressor gene function or gain of proto-oncogene activity are expected to contribute to most of the genetic alterations in cancers. Other genetic alterations leading to the ability of cancer cells to metastasize are also expected to play a prominent role after transformed cells have acquired a growth advantage over their normal counterpart.

Studies on colorectal carcinogenesis have provided remarkable evidence supporting the roles of both tumor suppressor genes and proto-oncogenes in the development of human cancers.[2,3] A great number of studies have provided information that suggest colorectal carcinogenesis consists of an accumulation of genetic defects that involve both activation of oncogenes, such as c-*myc*, and inactivation of tumor suppressor genes, such as *ras* and p53. The molecular-based approach to the study of colorectal cancer has bolstered many theories about carcinogenesis by demonstrating: 1) various genetic alterations are found in human colon cancer; 2) loss of specific genes implying that such genes have tumor suppressor function; 3) activation of proto-oncogenes by gene amplifications or point mutations. Although there is a myriad of genetic changes in human colorectal cancers, it is conceivable that in any individual cancer specimen only certain combinations of genetic alterations are sufficient to induce oncogenesis. In addition, the absolute sequence of genetic alterations may not be necessary; rather it is the cumulative effect of these alterations that is of major importance for oncogenesis (Fig. 2.7). Although the only shortcoming of the studies on genetic alterations in colon cancer is the difficulty in detecting changes in the expression of particular oncogenes, studies on colon cancer has by far provided the most thorough identification of genetic defects in a common human cancer.

5.2 GENETIC ALTERATIONS OF *myc*

Mutations affecting the c-*myc* gene exemplify the myriad of genetic alterations that can transform a proto-oncogene into a potentially lethal oncogene (Fig. 5.1). Virtually all mutations affecting c-*myc* are acquired, since no known heritable alteration of this gene has been reported.

5.2.1 Gene Amplification

Genetic alterations of the *myc* family of proto-oncogenes have been identified in a variety of human cancers; however, a pivotal role for *myc* in the pathogenesis of human cancers has only been strongly implicated in a limited number of cancers. For example, amplification of N-*myc* has clear prognostic significance in neuroblastoma.[4] A higher copy number of N-*myc* portends a worse prognosis.[5] In fact, N-*myc* amplification has been used as a tool to classify neuroblastomas in clinical therapeutic trials. Amplification

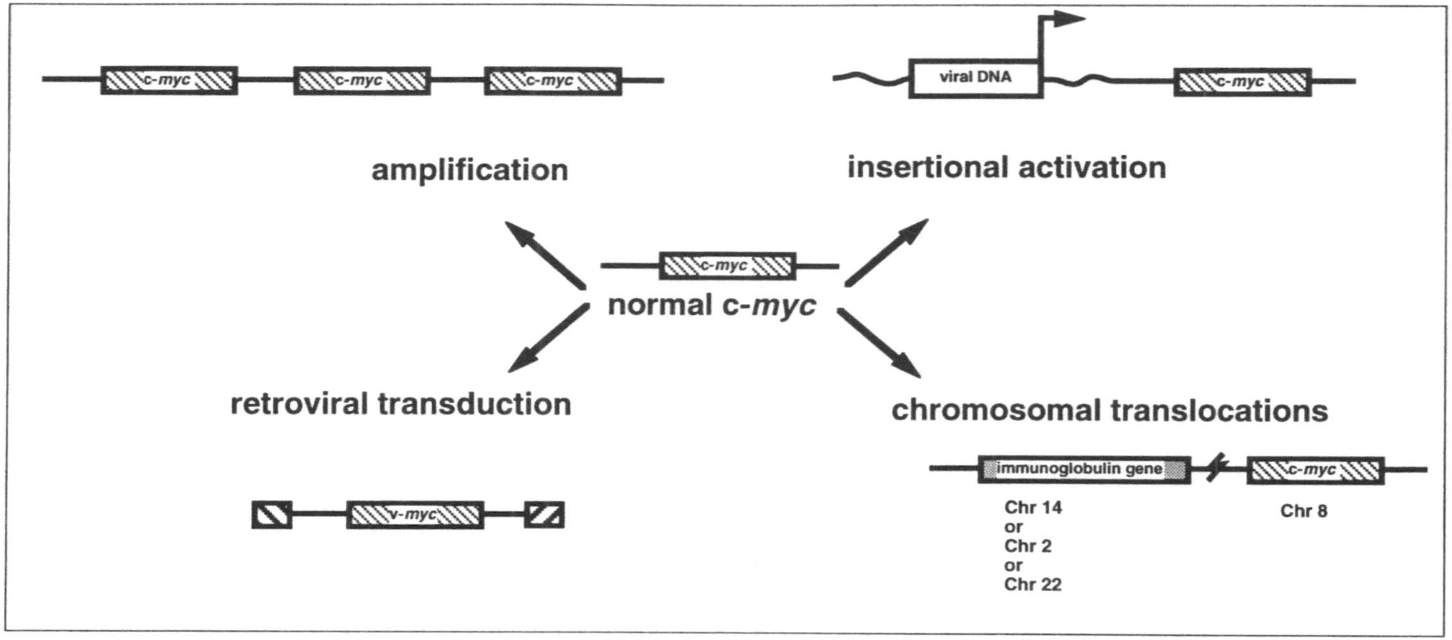

Fig. 5.1. Alterations of c-myc in neoplasia. The c-myc *gene undergoes virtually all known genetic alterations that activate its transcription: amplification, insertional activation, retroviral transduction and chromosomal translocation. Retroviral transduction has not been found in human diseases and insertional activation has not been conclusively documented in any human cancers. Overexpression of* c-myc*, which is found frequently in human breast and colon cancers, is not depicted since the genetic mechanism underlying this alteration is as yet unestablished. Translocation of* c-myc *from its normal position on human chromosome 8 to one of three immunoglobulin loci on chromosomes 2, 22 or 14 is indicated.*

of L-*myc* was a lesion that initially led to its identification in human small cell lung cancer by sequence homology to c-*myc*.[6] Amplification of c-*myc* has been detected in significant numbers in lung carcinoma, ovarian cancers as well as human breast carcinoma.

5.2.2 Chromosomal Translocation

One of the most recognized genetic lesions involving c-*myc* is its translocation to immunoglobulin light or heavy chain gene regulatory regions in B-cell malignancies, such as Burkitt's lymphoma and acute B-lymphocytic leukemia.[7-9] Reciprocal translocations of c-*myc* are probably critical pathogenetic events in these diseases.[10,11] Burkitt's lymphoma was initially identified as a prevalent early childhood aggressive B-cell cancer in East Africa.[12] The endemic African form of Burkitt's lymphoma frequently presents as a jaw mass and is associated with the Epstein-Barr virus (EBV). The sporadic form of Burkitt's lymphoma often presents as gastrointestinal cancers that are found in older children and young adults residing in Europe or the United States. Sporadic Burkitt's lymphoma is less often associated with EBV infection. In 90% of Burkitt's lymphoma, the c-*myc* locus on human chromosome 8q24 is often reciprocally translocated and juxtaposed to the immunoglobulin heavy chain locus on chromosome 14, to the light chain lambda locus on chromosome 22 or to the light chain kappa locus on chromosome 2 (Fig. 5.1). AIDS-related lymphomas, which are clinically intermediate and high-grade diseases, also commonly have alterations of c-*myc*.[13]

The most common chromosomal translocation type, t(8;14), is comprised of a breakpoint within the c-*myc* intron 1 or in the 5' untranscribed region with translocation of the entire *myc* region to the IgH locus, commonly in the switch region of μ, Sμ.[14,15] The consequence is a head-to-head transcriptional configuration for c-*myc* and IgH.[16,17] The translocated c-*myc* alleles are expressed from cryptic promoters in intron 1. With the less common variant translocations, t(2;8) and t(8;22), the breaks in c-*myc* are 3' of the transcribed region up to 140 kb downstream of c-*myc*.[18-23] Since most of the sites for translocation breakpoints are in the IgH switch or *V(D)J* join regions, it has been suggested that translocations are results of aberrant immunoglobulin genetic recombination events. In support of this notion are short sequences resembling heavy chain switch repeats located at the junctions of Ig and c-*myc*. Furthermore, repeats of nonamer/heptamer *V(D)J* joining sequences have been identified at breakpoints involving the D_H or J_H regions.

The enhancer of the immunoglobulin heavy chain locus (Eμ) is occasionally associated with the translocated c-*myc* allele in Burkitt's lymphoma.[14,15,24] Since Eμ is located in between J_H and Sμ, it is often translocated and associated with the upstream region of c-*myc* that does not include the coding region. As such, Eμ does not act in *cis* as an enhancer in the common Burkitt's lymphoma t(8;14) translocation. Several exceptions, such as translocations in Daudi, Manca and BL31 Burkitt's lymphoma cell lines, exist in which the Eμ enhancer may contribute to c-*myc* expression. Because of the infrequent juxtaposition of Ig enhancers to the c-*myc* locus, it has been proposed that the open chromatin structure within the immunoglobulin gene loci maintains the translocated c-*myc* allele in a transcriptionally active state. This hypothesis is in part supported by the observation that c-*myc* alleles are in inactive chromatin in mature B-cells.[25] More recent studies suggest that certain translocated c-*myc* alleles may be maintained in an active state by a juxtaposed immunoglobulin locus control region.[26]

5.2.3 Point Mutations in c-*myc*

Another intriguing consequence of being juxtaposed to the immunoglobulin loci is the susceptibility of the translocated c-*myc* allele to hypermutational events that normally operate on the V region (Fig. 5.2). Specifically, frequent mutations are found at the 3' end of exon 1 that appear to abolish transcriptional attenuation.[10,27] Such mutations are thought to increase transcriptional elongation resulting in elevated c-*myc* mRNA levels. Mutations within the coding sequence of c-*myc* have been detected in a limited number of Burkitt's lymphoma cell lines.[28] A mutation, Thr58→Ala, has been found not only in Burkitt's lymphoma but an equivalent mutation is also found in three isolates of v-*myc*.[29] This mutation appears to enhance the ability of c-*myc* to transform chicken cells and rodent fibroblasts. Moreover, Thr58 has been shown to be phosphorylated by mitogen activated protein kinase (MAP kinase). The exact mechanism of activation by this mutation is unknown, although a recent model suggests that a putative p107 mediated phosphorylation of Thr58 may downregulate c-Myc function (see chapter 7).

Recently, somatic mutations within the c-*myc* coding sequence have been detected in up to 60% of Burkitt's and AIDS-related lymphoma cases (Fig. 5.2).[30-34] The mutations tend to cluster within specific regions of exon 2. Mutations in exon 3 are rarer. Many of the mutations in exon 2 are localized to several phosphorylation sites that may affect the ability of c-*myc* to activate transcription. While these apparently nonrandom mutations may contribute to the activation of c-*myc*, the biological significance of these mutations has now emerged. These tumor-derived Myc mutations aug-

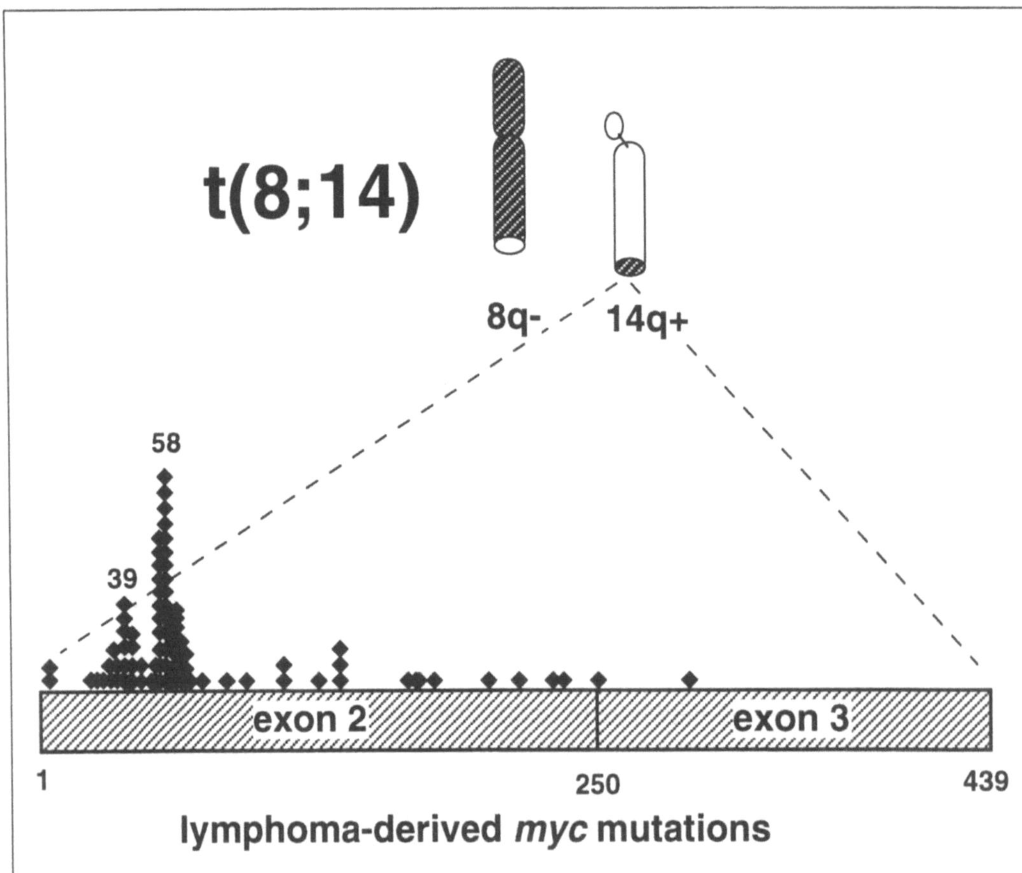

Fig. 5.2. Chromosomal translocation and coding sequence mutations in c-myc. The (8;14) translocation is shown at the top. Translocated alleles of c-myc in Burkitt's and AIDS related lymphomas are frequently mutated. Missense point mutations of myc *compiled from lymphoma-derived c-myc alleles and v-myc are indicated at the bottom as diamonds over the affected codons. The* myc *coding sequence is shown and labelled as exons 2 and 3. For v-myc mutations, the corresponding human c-myc positions are used to indicated the sites of alteration.*

ment transforming activity as well as cause a resistance of suppression of Myc transactivation by the retinoblastoma-related protein, p107 (see chapter 7).

5.3 ALTERATIONS OF *c-myc* AND ITS PRODUCT IN HUMAN CANCERS

Although specific genetic alterations of c-*myc* in examples of human malignancies are discussed above, there are myriads of studies implicating c-*myc* in a variety of cancers. Here, we will review alterations of c-*myc* in specific human cancers with an emphasis on the more frequent alterations in common solid tumors (excludes leukemia and lymphomas). This survey is meant to emphasize recurrent independent observations of specific alterations of c-*myc* in human cancers. In reviewing the literature, it becomes evident that the most easily detectable genetic changes, such as gene amplification or chromosomal translocation, are more frequently studied. The more difficult assays, such as measuring mRNA levels in primary tumor samples, are less frequently used and therefore rarely reported. The state-of-the-art, therefore, yields a view of genetic alterations of c-*myc* in human cancers skewed toward gene amplification.

5.3.1. Brain, Head and Neck Tumors

Human brain tumors are infrequently affected by alterations of c-*myc*, although in one study structural alterations were found in a case of glioblastoma multiforme.[35] On the other hand, loss of heterozygosity of *mxi*-1, which encodes a helix-loop-helix leucine zipper protein that counteracts the activity of c-Myc, has been found in over half of glioblastoma multiforme cases.[36] Amplification of N-*myc*, on the other hand, is essentially *sinequanon* for poor prognosis in neuroblastoma.[5] The role of c-*myc* in head and neck cancers is unclear. A minority (<10%) of head and neck cancers with poor prognosis contain c-*myc* amplification, whereas about one-quarter of squamous cell carcinoma cell lines have amplified c-*myc*.[37-39] These observations suggest that structural alterations of the c-*myc* gene are found in human head and neck, and brain cancers, although at low frequencies.

5.3.2. Lung Cancer

Lung cancer, a leading cause of death, consists of essentially three major subgroups: small cell (SCLC), squamous cell and adenocarcinoma. The most aggressive form, small cell carcinoma of the lung, was known to contain homogenously staining regions (HSRs) in chromosomes as well as double-minute (DM) chromosomes.[40] Subsequently, c-*myc* and N-*myc* were thought to be amplified in these cancers,[41-47] manifesting as HSRs or DMs; however, additional studies revealed a novel *myc* gene termed L-*myc* that is amplified in a subset of lung cancers.[6] On rare occasions, L-*myc* amplification is associated with an intrachromosomal rearrangement of L-*myc* that generates a fusion gene termed *rlf*-L-*myc*.[48,49] Albeit rare, the recurrence of the *rlf*-L-*myc* fusion gene in lung cancer suggests a role for this fusion gene in the genesis of these cancers. c-*myc* itself is also amplified in a minority of lung cancers. Overall, an increase in the expression of one of the related *myc* genes occurs in about 10% to 40% of small cell lung cases or cell lines. Since amplification of *myc* is more frequently found in cell lines than primary tumor samples, it was surmised that amplification signifies tumor progression. However, studies did not support this contention; rather they suggest that *myc* amplification in SCLC results from prior chemotherapy.[50] Thus, in human lung cancers, amplification of *myc* genes can be detected in about one-third of cases, suggesting that *myc* gene expression contributes to lung carcinogenesis.

5.3.3. Breast Cancer

Breast cancer, a leading cause of death of American women, displays chromosomal abnormalities that frequently involve amplification of 8q24, where c-*myc* is located.[51,52] Amplification of the c-*myc* gene was originally observed in breast cancer cell lines and has since been detected in 4% to 41% of primary breast tumor cases.[51,53-69] The possibility of a connection between deletion of the short arm of chromosome 1 and gene amplification was investigated.[70] One hundred twenty-two human primary breast tumors were examined for loss of heterozygosity (LOH) on 1p32-pter and for amplification of c-*myc*. LOH on 1p32-pter was observed in 57 (47%) of 122 informative tumors, and c-*myc* was amplified in 23 of the tumors. The significant correlation of loss of heterozygosity on chromosome 1p32-pter with amplification of the c-*myc* proto-oncogene suggests that telomeric sequences of 1p harbors an unidentified tumor suppressor gene(s), whose inactivation may be involved in c-*myc* family gene amplification.

Whereas rearrangement of c-*myc* occurs in less than 15% of breast carcinomas,[55,63] its overexpression is found in 30% to 40% of cases.[63,68,71-74] It is intriguing that c-*myc* expression is augmented by exposure of breast cancer cells to estrogen, and that lower c-*myc* expression is found in estrogen receptor-positive tumors treated preoperatively with the anti-estrogen, tamoxifen.[75] Studies of the clinical significance of c-*myc* amplification have produced inconclusive results, since amplification has been inconsistently correlated with nodal metastasis, poor survival, progesterone receptor negativity, histology and high cathepsin D expression.[54,59,69,74] The detection of elevated c-Myc protein levels in breast cancers is also of unclear significance, since fibrocystic breast disease appears to have increased c-Myc protein expression as well.[76,77] Studies that accumulated thus far strongly suggest the involvement of c-*myc* gene deregulation in pathogenesis of human breast cancer. Although the use of c-*myc* amplification as a prognostic marker has been touted by several studies, its clinical application remains controversial.[68,69,74]

5.3.4. Colorectal Cancer

Perhaps the most compelling and elegant epic of molecular pathogenesis of human cancer is that emerging from the study of colorectal carcinomas.[78,79] The story of c-*myc* in colorectal carcinogenesis and gastrointestinal malignancies is as yet incomplete and remains intriguing. High levels of c-myc expression were detected in colorectal polyps, and the transcript levels significantly correlated with polyp size and dysplastic histology.[80,81] c-*myc* mRNA was also found to be elevated in about two-thirds of the polyps examined demonstrating that premalignant lesions overexpress the gene. N-*myc* and L-*myc* transcripts can be detected at low abundance in normal colonic mucosa and found to be frequently, but only modestly overexpressed in tumors and polyps.[82] Elevated c-Myc protein expression in polyps has been observed as well.[83,84] Adenomatous polyps revealed the most intense pattern of c-Myc staining in dysplastic areas. These studies suggest that colonic polyps frequently have elevated c-*myc* expression. Presumably the expression of Myc contributes to the hyperplastic state of these deranged premalignant cells.

Expression of the c-*myc* gene is elevated (5-fold to 40-fold over normal) in two-thirds of sporadic human colorectal cancers, but amplification of c-*myc* is rare.[85-87] In contrast, gastric cancers rarely display alterations of the c-*myc* gene or its expression.[81,88,89] In one study of colorectal carcinomas, modest amplification of c-*myc* was found in 53.8% (7/13) mucinous tumors and 42.3% (3/7) poorly differentiated tumors.[90] This contrasts with amplification in 6.9% (2/29) moderately to well differentiated tumors, although these findings have not been independently verified. Overexpression of c-*myc* in colorectal cancers, however, has been reported in several independent studies. In fact, the mechanism underlying c-*myc* overexpression in colorectal cancers has been investigated. One study demonstrated that the degree of hypomethylation of the c-*myc* locus is directly proportional to c-*myc* expression and the clinical stage of the tumors, with metastatic lesions containing the most hypomethylated alleles.[91]

An alternative hypothesis that these tumors are defective in a gene necessary for the regulation of c-*myc* expression was tested by fusing an osteosarcoma cell line that exhibits normal c-*myc* regulation with two colon carcinoma cell lines that express deregulated levels of c-*myc* mRNA.[92] The levels of c-*myc* mRNA in all of the hybrid clones were normal suggesting that loss of function of a *trans*-acting regulator is responsible for the deregulation of c-*myc* expression. Analysis of tumor and normal tissue pairs from patients with primary colorectal lesions indicated that deregulation of c-*myc* expression in the tumors correlates with frequent (9/19) loss of alleles of syntenic markers on chromosome 5q.

The deregulated expression of c-*myc* in a colorectal cancer cell line is suppressed by fusion with a cell that regulates c-*myc* normally. In addition, transfer of chromosome 5 by microcell fusion resulted in suppression of deregulated expression in cells that were no longer tumorigenic in nude mice.[93] Loss of the transferred chromosome resulted in reexpression of the tumorigenic phenotype and in constitutive elevated expression of c-*myc*. These results suggest that loss of function of a tumor-suppressor gene on chromosome 5 results in deregulated c-*myc* expression and is a necessary, but probably not sufficient, event for the expression of the tumorigenic phenotype in a subset of colon carcinomas. The identity of the suppressor of c-*myc* expression on chromosome 5 remains unknown; however, it seems reasonable to hypothesize that the familial adenomatous polyposis gene on 5q may be associated with the regulation of c-*myc* expression.

5.3.5. Gynecologic Tumors

The c-*myc* gene is frequently altered in certain gynecological tumors, except vulvar carcinomas.[94,95] In one study, 35 human tumors of the uterine cervix were evaluated for alterations of c-*myc*.[96] Amplification and/or rearrangement of the c-*myc* oncogene were detected in most (approximately 90%) samples (48% showed amplification and 43% presented both alterations), which were mostly stage II cervical carcinomas. Invasive carcinomas of the uterine cervix of 38 patients were examined in another study for the presence of human papillomavirus (HPV) genomes and for alterations of the c-*myc* and Ha-*ras* oncogenes.[97] Southern blot and polymerase chain reaction analyses revealed the presence of the genome of HPV type 16 in 17 tumors (45%), that of HPV type 18 in 3 tumors (8%), and that of unknown types in 16 others (42%). c-*myc* amplification was found in only 1 of 38 tumors, while there was no Ha-ras amplification. Overexpression of the c-*myc* gene, however, was observed in 15 (44%) of the 34 tumors analyzed, while there was no overexpression of Ha-*ras*. Relapse-free survival rates of the 23 squamous cell carcinomas analyzed at 24 months were 55% in tumors with c-*myc* overexpression and 100% in case of tumors with no c-*myc* overexpression, respectively.

The expression of the c-Myc protein has also been studied in paraffin-embedded specimens of cervical biopsies using a monoclonal anti-c-Myc antibody.[98] There was no correlation between the clinical stage of the disease at presentation and c-Myc expression. Fifteen patients with c-Myc negative cervical cancers had better disease free (mean of 95.4 months) and total survival (mean of 118.0 months) as compared to the 16 patients with c-Myc positive tumors (28.4 and 48.4 months, respectively). The c-Myc positive tumors were more likely to recur as extra-pelvic metastatic disease. These studies taken together suggest that the c-*myc* status of cervical cancer offers a prognostic indicator that could be useful in guiding treatment decisions.

Amplification of the c-*myc* oncogene correlated with advanced-stage poorly differentiated primary endometrial adenocarcinomas in one small series.[99] Eleven of sixteen tumor samples, which were analyzed for c-*myc* and c-*neu* oncogene alterations, exhibited amplification of the *neu* gene. Four of these eleven patients died at an average of 16 months after diagnosis. Ten of fifteen tumor samples exhibited amplification of the c-*myc* gene. Five of the ten patients died of disease at an average of 13.4 months after diagnosis. The remaining five are free of disease after a mean follow period of 31.2 months. Six of the sixteen tumor specimens exhibited amplification of the both c-*neu* and c-*myc* genes, and four of these patients died of recurrent disease, suggesting that oncogene amplification may predict biologically aggressive adenocarcinomas of the endometrium.

c-*myc* amplification appears to be a common finding in advanced-stage ovarian cancer (17% to 30% of cases); however clinical correlations do not suggest a prognostic role for c-myc amplification in primary ovarian cancer.[100-104] In an archival polymerase chain reaction study, 17% of 30 ovarian carcinoma cases displayed an increase in c-*myc* copy number.[102] In a study of primary tumors, DNAs from 24 ovarian tumors, including 16 carcinomas, were assayed for amplification of c-*myc*.[101] Protooncogene amplification was not detected in the DNA of benign ovarian neoplasms, or of ovarian carcinomas with low malignant potential. Amplification of c-*myc*, however, was detected in 6 of 12 (50%) cases of invasive carcinoma. c-*myc* amplification was frequently associated with morphologic nuclear anaplasia and high mitotic count. While the total number of cases in this study was limited, the trend of the results suggests that c-*myc* amplification may be involved in the pathogenesis of aggressive common epithelial tumors of the ovary. In another study, 5 of 17(29.4%) tumor samples demonstrated amplification of the c-*myc*.[100] The five patients with these tumors exhibited a median survival of 17 months. Of the 12 patients without evidence of tumor-associated c-*myc* amplification, 5 had disease-free survival for an average of 36.8 months. The remaining seven patients, the majority of whom had poorly differentiated lesions with a normal c-*myc* copy number, had a median survival of 9 months. There was no correlation between c-*myc* amplification, grade of tumor differentiation or response to platinol-based chemotherapy.

5.3.6. Male Urologic Tumors

Cancers of the male reproductive system, such as testicular cancer have been infrequently associated with altered c-Myc expression.[105] Genetic alteration of c-*myc* is virtually absent in testicular cancers. In contrast, elevated c-*myc* expression has been observed in human prostate cancers.[106-111] In fact, experimental prostatic carcinoma in the rat can be elicited by deregulated expression of *ras* and v-*myc*.[112,113] In humans, about one-quarter of prostatic carcinomas are found to have elevated c-*myc* expression. More importantly, however, about one-third of prostatic carcinoma cases are associated with LOH involving *mxi*-1 (see chapter 8; Wechsler and Dang, unpublished observation). These results suggest that the Myc/Max protein network may play a central role in the pathogenesis of human prostatic carcinoma.

Summary

Through reviewing the different organ systems, it seems apparent that the c-*myc* oncogene has received its fair share of attention in the pathogenesis of human cancers. In breast, colon and ovarian cancers, significant fractions of cases appear to have alterations of the c-*myc* gene or its expression. In lung cancers, amplification of L-*myc* as well as c-*myc* have been detected in less than half of the cases. The clinical implications of these *myc* alterations are generally unknown with the emphatic exception of N-*myc* amplification in neuroblastomas. High N-*myc* copy number portends high grade tumors with poorer prognosis; this observation has now become a practice in predicting the clinical outcome of neuroblastomas. Although this achievement may be perceived as a small step forward, it is certainly a significant leap from the bench to the bedside.

5.4 DEREGULATED EXPRESSION OF *myc* CAUSES NEOPLASTIC TRANSFORMATION AND ABERRANT DEVELOPMENT IN TRANSGENIC ANIMALS

It is well-established that alterations of the c-*myc* gene correlate well with a variety of human cancers. In certain cases, such as Burkitt's lymphoma, the role of c-*myc* in the pathogenesis of the cancer is almost beyond any reasonable doubt. Nonetheless, it is virtually impossible to fulfill Koch's postulate, which requires the introduction of the causative agent back into the animal to recreate the disease to implicate any oncogene in the genesis of human cancers. To support the hypothesis that deregulated expression of oncogenes contributes to neoplasia, transgenic animals have been created in which *myc* gene expression is deregulated in specific tissues. Readers who are interested in the role of transgenic animals in the study of cancers should seek more comprehensive reviews.[114-116]

As in the case of human Burkitt's lymphoma, in which c-*myc* expression is enforced, early elegant transgenesis studies showed that deregulated expression of *myc* in lymphoid tissues result in lymphomagenesis.[117-120] The time course of lymphoma development suggests that multiple additional genetic alterations are required for the onset of the lymphomas.[121] Subsequent studies indicated that polyclonal lymphoid hyperplasia precedes the onset of frank lymphomas in these transgenic animals.[121] In addition, the cooperativity of c-*myc* with activated *ras* and other oncogenes has been well established in transgenic animals.[122-130] Ectopic transgenic expression of c-*myc* in various tissues predispose them to malignant transformation.[125,131-141] These landmark studies have critically supported the hypothesis that deregulated expression of c-*myc* is able to induce neoplastic transformation in animals. Furthermore, they underscore the importance of cooperativity between oncogenes.

References

1. Modrich P. Mismatch repair, genetic stability, and cancer. [Review]. Science 1994; 266:1959-60.
2. Fearon ER, Vogelstein B. A genetic model for colorectal tumorigenesis. [Review]. Cell 1990; 61:759-67.
3. Jiricny J. Colon cancer and DNA repair: have mismatches met their match?. [Review]. Trends Genet 1994; 10:164-8.
4. Schwab M, Varmus HE, Bishop JM, et al. Chromosome localization in normal human cells and neuroblastomas of a gene related to c-myc. Nature 1984; 308:288-91.
5. Brodeur GM, Seeger RC, Schwab M, Varmus HE, Bishop JM. Amplification of N-myc in untreated human neuroblastomas correlates with advanced disease stage. Science 1984; 224:1121-4.
6. Nau MM, Brooks BJ, Battey J, et al. L-myc, a new myc-related gene amplified and expressed in human small cell lung cancer. Nature 1985; 318:69-73.
7. Nowell PC, Croce CM. Chromosomal approaches to the molecular basis of neoplasia. [Review]. Symp Fundam Cancer Res 1986; 39:17-29.
8. Leder P, Battey J, Lenoir G, et al. Translocations among antibody genes in human cancer. [Review]. Science 1983; 222:765-71.
9. Guerrasio A, Avanzi GC, Pegoraro L, et al. Rearrangement of the c-myc oncogene with heavy-chain immunoglobulin enhancer in tumor DNA from an acute lymphoblastic leukemia patient. J Natl Cancer Inst 1987; 78:845-51.
10. Pelicci PG, Knowles Dd, Magrath I, Dalla-Favera R. Chromosomal breakpoints and structural alterations of the c-myc locus differ in endemic and sporadic forms of Burkitt lymphoma. Proc Natl Acad Sci USA 1986; 83:2984-8.

11. Klein G. Constitutive activation of oncogenes by chromosomal translocations in B-cell derived tumors. [Review]. Aids Res 1986; 2:S167-76.
12. Magrath I. The pathogenesis of Burkitt's lymphoma. [Review]. Adv Cancer Res 1990; 55:133-270.
13. Gaidano G, Pastore C, Lanza C, Mazza U, Saglio G. Molecular pathology of AIDS-related lymphomas. Biologic aspects and clinicopathologic heterogeneity. [Review]. Ann Hematol 1994; 69:281-90.
14. Marcu K, Bossone S, Patel A. Myc function and regulation. Ann Rev Biochem 1992; 61:809-860.
15. Spencer CA, Groudine M. Control of c-myc regulation in normal and neoplastic cells. [Review]. Adv Cancer Res 1991; 56:1-48.
16. Taub R, Kirsch I, Morton C, et al. Translocation of the c-myc gene into the immunoglobulin heavy chain locus in human Burkitt lymphoma and murine plasmacytoma cells. Proc Natl Acad Sci USA 1982; 79:7837-41.
17. Hamlyn PH, Rabbitts TH. Translocation joins c-myc and immunoglobulin gamma 1 genes in a Burkitt lymphoma revealing a third exon in the c-myc oncogene. Nature 1983; 304:135-9.
18. Croce CM, Thierfelder W, Erikson J, et al. Transcriptional activation of an unrearranged and untranslocated c-myc oncogene by translocation of a C lambda locus in Burkitt. Proc Natl Acad Sci USA 1983; 80:6922-6.
19. Erikson J, Nishikura K, ar-Rushdi A, et al. Translocation of an immunoglobulin kappa locus to a region 3' of an unrearranged c-myc oncogene enhances c-myc transcription. Proc Natl Acad Sci USA 1983; 80:7581-5.
20. Graham M, Adams JM. Chromosome 8 breakpoint far 3' of the c-myc oncogene in a Burkitt's lymphoma 2;8 variant translocation is equivalent to the murine pvt-1 locus. EMBO J 1986; 5:2845-51.
21. Henglein B, Synovzik H, Groitl P, Bornkamm GW, Hartl P, Lipp M. Three breakpoints of variant t(2;8) translocations in Burkitt's lymphoma cells fall within a region 140 kilobases distal from c-myc. Mol Cell Biol 1989; 9:2105-13.
22. Sun LK, Showe LC, Croce CM. Analysis of the 3' flanking region of the human c-myc gene in lymphomas with the t(8;22) and t(2;8) chromosomal translocations. Nucleic Acids Res 1986; 14:4037-50.
23. Kornblau SM, Goodacre A, Cabanillas F. Chromosomal abnormalities in adult non-endemic Burkitt's lymphoma and leukemia: 22 new reports and a review of 148 cases from the literature. [Review]. Hematol Oncol 1991; 9:63-78.
24. Rabbitts TH, Forster A, Baer R, Hamlyn PH. Transcription enhancer identified near the human Cμ immunoglobulin heavy chain gene is unavailable to the translocated c-myc gene in a Burkitt lymphoma. Nature 1983; 306:806-9.

25. Dyson PJ, Littlewood TD, Forster A, Rabbitts TH. Chromatin structure of transcriptionally active and inactive human c-myc alleles. EMBO J 1985; 4:2885-91.
26. Madisen L, Groudine M. Identification of a locus control region in the immunoglobulin heavy-chain locus that deregulates c-myc expression in plasmacytoma and Burkitt's lymphoma cells. Genes Dev 1994; 8:2212-26.
27. Eick D, Polack A, Kofler E, Bornkamm GW. The block of elongation in c-myc exon 1 is abolished in Burkitt's lymphoma cell lines with variant translocation. Oncogene 1988; 3:397-403.
28. Rabbitts T, Hamlyn P, Baer R. Altered nucleotide sequences of a translocated c-*myc* gene in Burkitt lymphoma. Nature 1983; 306:760-765.
29. Papas T, Lautenberg J. Sequence curiosity in v-*myc* oncogene. Nature 1985; 318:237.
30. Yano T, Sander C, Clark H, Dolezal M, Jaffe E, Raffeld M. Clustered mutations in the second exon of the MYC gene in sporadic Burkitt's lymphoma. Oncogene 1993; 8:2741-2748.
31. Clark HM, Yano T, Otsuki T, Jaffe ES, Shibata D, Raffeld M. Mutations in the coding region of c-myc in aids-associated and other aggressive lymphomas. Cancer Res 1994; 54:3383-3386.
32. Bhatia K, Huppi K, Spangler G, Siwarski D, Iyer R, Magrath I. Point mutations in the c-Myc transactivation domain are common in Burkitt's lymphoma and mouse plasmacytoma. Nature Genet 1993; 5:56-61.
33. Bhatia K, Spangler G, Gaidano G, Hamdy N, Dalla-Favera R, Magrath I. Mutations in the coding region of c-myc occur frequently in acquired immunodeficiency syndrome-associated lymphomas. Blood 1994; 84:883-8.
34. Albert T, Urlbauer B, Kohlhuber F, Eick D, Hammersen B. Ongoing mutations in the n-terminal domain of c-myc affect transactivation in burkitts lymphoma cell lines. Oncogene 1994; 9:759-763.
35. Trent J, Meltzer P, Rosenblum M, et al. Evidence for rearrangement, amplification, and expression of c-myc in a human glioblastoma. Proc Natl Acad Sci USA 1986; 83:470-3.
36. Albarosa R, Didonato S, Finocchiaro G. Redefinition of the coding sequence of the MXI1 gene and identification of a polymorphic repeat in the 3' non-coding region that allows the detection of loss of heterozygosity of chromosome 10q25 in glioblastomas. Human Genet 1995; 95:709-711.
37. Yin XY, Donovan-Peluso M, Whiteside TL, et al. Gene amplification and gene dosage in cell lines derived from squamous cell carcinoma of the head and neck [see comments]. Genes Chrom Can 1991; 3:443-54.
38. Merritt WD, Weissler MC, Turk BF, Gilmer TM. Oncogene amplification in squamous cell carcinoma of the head and neck. Archives of Otolaryngology Head Neck Surg 1990; 116:1394-8.

39. Field JK. Oncogenes and tumour-suppressor genes in squamous cell carcinoma of the head and neck. [Review]. Eur J Cancer B Oral Oncol 1992; 1:67-76.
40. Makela TP, Saksela K, Alitalo K. Amplification and rearrangement of L-myc in human small-cell lung cancer. [Review]. Mut Res 1992; 276:307-15.
41. Little CD, Nau MM, Carney DN, Gazdar AF, Minna JD. Amplification and expression of the c-myc oncogene in human lung cancer cell lines. Nature 1983; 306:194-6.
42. Nau MM, Carney DN, Battey J, et al. Amplification, expression, and rearrangement of c-myc and N-myc oncogenes in human lung cancer. Curr Top Microbiol Immunol 1984; 113:172-7.
43. Vastrik I, Makela TP, Koskinen PJ, Saksela K, Alitalo K. myc, max, and a novel rlf-L-myc fusion protein in small-cell lung cancer. [Review]. Princess Takamatsu Symp 1991; 22:307-18.
44. Gazdar AF. The molecular and cellular basis of human lung cancer. [Review]. Anticancer Res 1994; 14:261-7.
45. Bergh JC. Gene amplification in human lung cancer. The myc family genes and other proto-oncogenes and growth factor genes. [Review]. Amer Rev Resp Dis 1990; 142:S20-6.
46. Birrer MJ, Minna JD. Genetic changes in the pathogenesis of lung cancer. [Review]. Ann Rev Med 1989; 40:305-17.
47. Wong AJ, Ruppert JM, Eggleston J, Hamilton SR, Baylin SB, Vogelstein B. Gene amplification of c-myc and N-myc in small cell carcinoma of the lung. Science 1986; 233:461-4.
48. Makela TP, Saksela K, Evan G, Alitalo K. A fusion protein formed by L-myc and a novel gene in SCLC. EMBO J 1991; 10:1331-5.
49. Makela TP, Kere J, Winqvist R, Alitalo K. Intrachromosomal rearrangements fusing L-myc and rlf in small-cell lung cancer. Molecular & Cellular Biology 1991; 11:4015-21.
50. Johnson BE, Ihde DC, Makuch RW, et al. myc family oncogene amplification in tumor cell lines established from small cell lung cancer patients and its relationship to clinical status and course. J Clin Invest 1987; 79:1629-34.
51. Saint-Ruf C, Gerbault-Seureau M, Viegas-Pequignot E, Zafrani B, Cassingena R, Dutrillaux B. Proto-oncogene amplification and homogeneously staining regions in human breast carcinomas. Genes Chrom Can 1990; 2:18-26.
52. Kallioniemi A, Kallioniemi OP, Piper J, et al. Detection and mapping of amplified DNA sequences in breast cancer by comparative genomic hybridization. Proc Natl Acad Sci USA 1994; 91:2156-60.
53. McKenzie SJ. Diagnostic utility of oncogenes and their products in human cancer. [Review]. Biochim Biophys Acta 1991; 1072:193-214.
54. Adnane J, Gaudray P, Simon MP, Simony-Lafontaine J, Jeanteur P, Theillet C. Proto-oncogene amplification and human breast tumor phenotype. Oncogene 1989; 4:1389-95.

55. Bonilla M, Ramirez M, Lopez-Cueto J, Gariglio P. In vivo amplification and rearrangement of c-myc oncogene in human breast tumors. J Natl Cancer Inst 1988; 80:665-71.
56. Borg A, Baldetorp B, Ferno M, Olsson H, Sigurdsson H. c-myc amplification is an independent prognostic factor in postmenopausal breast cancer. Int J Can 1992; 51:687-91.
57. Champeme MH, Bieche I, Hacene K, Lidereau R. Oncogene amplification per se: an independent prognostic factor in human breast cancer. Mol Carcinog 1994; 11:189-91.
58. Cline MJ, Battifora H, Yokota J. Proto-oncogene abnormalities in human breast cancer: correlations with anatomic features and clinical course of disease. J Clin Onc 1987; 5:999-1006.
59. Garcia I, Dietrich PY, Aapro M, Vauthier G, Vadas L, Engel E. Genetic alterations of c-myc, c-erbB-2, and c-Ha-ras protooncogenes and clinical associations in human breast carcinomas. Can Res 1989; 49:6675-9.
60. Kozbor D, Croce CM. Amplification of the c-myc oncogene in one of five human breast carcinoma cell lines. Cancer Res 1984; 44:438-41.
61. Machotka SV, Garrett CT, Schwartz AM, Callahan R. Amplification of the proto-oncogenes int-2, c-erb B-2 and c-myc in human breast cancer. Clin Chim Acta 1989; 184:207-17.
62. Mariani-Costantini R, Escot C, Theillet C, et al. In situ c-myc expression and genomic status of the c-myc locus in infiltrating ductal carcinomas of the breast. Cancer Res 1988; 48:199-205.
63. Munzel P, Marx D, Kochel H, Schauer A, Bock KW. Genomic alterations of the c-myc protooncogene in relation to the overexpression of c-erbB2 and Ki-67 in human breast and cervix carcinomas. J Cancer Res Clin Oncol 1991; 117:603-7.
64. Pavelic ZP, Steele P, Preisler HD. Evaluation of c-myc proto-oncogene in primary human breast carcinomas. Anticancer Res 1991; 11:1421-7.
65. Seshadri R, Matthews C, Dobrovic A, Horsfall DJ. The significance of oncogene amplification in primary breast cancer. [Review]. Int J Can 1989; 43:270-2.
66. Tang RP, Kacinski B, Validire P, et al. Oncogene amplification correlates with dense lymphocyte infiltration in human breast cancers: a role for hematopoietic growth factor release by tumor cells? J Cell Biochem 1990; 44:189-98.
67. Tsuda H, Hirohashi S, Hirota T, Shimosato Y. Alterations in copy number of c-erbB-2 and c-myc proto-oncogenes in advanced stage of human breast cancer. Acta Pathol Jpn 1991; 41:19-23.
68. Tripathy D, Benz CC. Activated oncogenes and putative tumor suppressor genes involved in human breast cancers. [Review]. Can Treat Res 1992; 63:15-60.
69. Tavassoli M, Quirke P, Farzaneh F, Lock NJ, Mayne LV, Kirkham N. c-erbB-2/c-erbA co-amplification indicative of lymph node metastasis, and c-myc amplification of high tumour grade, in human breast carcinoma. Br J Can 1989; 60:505-10.

70. Bieche I, Champeme MH, Lidereau R. A tumor suppressor gene on chromosome 1p32-pter controls the amplification of MYC family genes in breast cancer. Cancer Res 1994; 54:4274-6.
71. Guerin M, Barrois M, Terrier MJ, Spielmann M, Riou G. Overexpression of either c-myc or c-erbB-2/neu proto-oncogenes in human breast carcinomas: correlation with poor prognosis. Oncogene Res 1988; 3:21-31.
72. Madsen MW, Lykkesfeldt AE, Laursen I, Nielsen KV, Briand P. Altered gene expression of c-myc, epidermal growth factor receptor, transforming growth factor-alpha, and c-erb-B2 in an immortalized human breast epithelial cell line, HMT-3522, is associated with decreased growth factor requirements. Cancer Res 1992; 52:1210-7.
73. Benz CC, Scott GK, Santos GF, Smith HS. Expression of c-myc, c-Ha-ras1, and c-erbB-2 proto-oncogenes in normal and malignant human breast epithelial cells. JNCI 1989; 81:1704-9.
74. Escot C, Theillet C, Lidereau R, et al. Genetic alteration of the c-myc protooncogene (MYC) in human primary breast carcinomas. Proc Natl Acad Sci USA 1986; 83:4834-8.
75. Le Roy X, Escot C, Brouillet JP, et al. Decrease of c-erbB-2 and c-myc RNA levels in tamoxifen-treated breast cancer. Oncogene 1991; 6:431-7.
76. Agnantis NJ, Mahera H, Maounis N, Spandidos DA. Immunohistochemical study of ras and myc oncoproteins in apocrine breast lesions with and without papillomatosis. Eur J Gyn Onc 1992; 13:309-15.
77. Spandidos DA, Yiagnisis M, Papadimitriou K, Field JK. ras, c-myc and c-erbB-2 oncoproteins in human breast cancer. Anticancer Res 1989; 9:1385-93.
78. Hamilton SR. Molecular genetics of colorectal carcinoma. [Review]. Cancer 1992; 70:1216-21.
79. Hamilton SR. Molecular genetic alterations as potential prognostic indicators in colorectal carcinoma. [Review]. Cancer 1992; 69:1589-91.
80. Sugio K, Kurata S, Sasaki M, Soejima J, Sasazuki T. Differential expression of c-myc gene and c-fos gene in premalignant and malignant tissues from patients with familial polyposis coli. Cancer Res 1988; 48:4855-61.
81. Imaseki H, Hayashi H, Taira M, et al. Expression of c-myc oncogene in colorectal polyps as a biological marker for monitoring malignant potential. Cancer 1989; 64:704-9.
82. Finley GG, Schulz NT, Hill SA, Geiser JR, Pipas JM, Meisler AI. Expression of the myc gene family in different stages of human colorectal cancer. Oncogene 1989; 4:963-71.
83. Watson JV, Stewart J, Cox H, Sikora K, Evan GI. Flow cytometric quantitation of the c-myc oncoprotein in archival neoplastic biopsies of the colon. Mol Cell Probes 1987; 1:151-7.

84. Stewart J, Evan G, Watson J, Sikora K. Detection of the c-myc oncogene product in colonic polyps and carcinomas. Br J Cancer 1986; 53:1-6.
85. Erisman MD, Rothberg PG, Diehl RE, Morse CC, Spandorfer JM, Astrin SM. Deregulation of c-myc gene expression in human colon carcinoma is not accompanied by amplification or rearrangement of the gene. Mol Cell Biol 1985; 5:1969-76.
86. Erisman MD, Scott JK, Watt RA, Astrin SM. The c-myc protein is constitutively expressed at elevated levels in colorectal carcinoma cell lines. Oncogene 1988; 2:367-78.
87. Yoshimoto K, Hirohashi S, Sekiya T. Increased expression of the c-myc gene without gene amplification in human lung cancer and colon cancer cell lines. Jpn J Cancer Res 1986; 77:540-5.
88. Nomura N, Yamamoto T, Toyoshima K, et al. DNA amplification of the c-myc and c-erbB-1 genes in a human stomach cancer. Jpn J Cancer Res 1986; 77:1188-92.
89. Tsuchiya T, Ueyama Y, Tamaoki N, Yamaguchi S, Shibuya M. Co-amplification of c-myc and c-erbB-2 oncogenes in a poorly differentiated human gastric cancer. Jpn J Cancer Res 1989; 80:920-3.
90. Heerdt BG, Molinas S, Deitch D, Augenlicht LH. Aggressive subtypes of human colorectal tumors frequently exhibit amplification of the c-myc gene. Oncogene 1991; 6:125-9.
91. Sharrard RM, Royds JA, Rogers S, Shorthouse AJ. Patterns of methylation of the c-myc gene in human colorectal cancer progression. Br J Cancer 1992; 65:667-72.
92. Erisman MD, Scott JK, Astrin SM. Evidence that the familial adenomatous polyposis gene is involved in a subset of colon cancers with a complementable defect in c-myc regulation. Proc Natl Acad Sci USA 1989; 86:4264-8.
93. Rodriguez-Alfageme C, Stanbridge EJ, Astrin SM. Suppression of deregulated c-MYC expression in human colon carcinoma cells by chromosome 5 transfer. Proc Natl Acad Sci USA 1992; 89:1482-6.
94. Milde-Langosch K, Becker G, Loning T. Human papillomavirus and c-myc/c-erbB2 in uterine and vulvar lesions. Virchows Arch A Pathol Anat Histopathol 1991; 419:479-85.
95. Tate JE, Mutter GL, Prasad CJ, Berkowitz R, Goodman H, Crum CP. Analysis of HPV-positive and -negative vulvar carcinomas for alterations in c-myc, Ha-, Ki-, and N-ras genes. Gynecol Oncol 1994; 53:78-83.
96. Ocadiz R, Sauceda R, Cruz M, Graef AM, Gariglio P. High correlation between molecular alterations of the c-myc oncogene and carcinoma of the uterine cervix. Cancer Res 1987; 47:4173-7.
97. Iwasaka T, Yokoyama M, Oh-uchida M, et al. Detection of human papillomavirus genome and analysis of expression of c-myc and Ha-ras oncogenes in invasive cervical carcinomas. Gynecol Oncol 1992; 46:298-303.

98. Sowani A, Ong G, Dische S, et al. c-myc oncogene expression and clinical outcome in carcinoma of the cervix. Mol Cell Probes 1989; 3:117-23.
99. Borst MP, Baker VV, Dixon D, Hatch KD, Shingleton HM, Miller DM. Oncogene alterations in endometrial carcinoma. Gynecol Oncol 1990; 38:364-6.
100. Baker VV, Borst MP, Dixon D, Hatch KD, Shingleton HM, Miller D. c-myc amplification in ovarian cancer. Gynecol Oncol 1990; 38:340-2.
101. Sasano H, Garrett CT, Wilkinson DS, Silverberg S, Comerford J, Hyde J. Protooncogene amplification and tumor ploidy in human ovarian neoplasms. Hum Pathol 1990; 21:382-91.
102. Schreiber G, Dubeau L. C-myc proto-oncogene amplification detected by polymerase chain reaction in archival human ovarian carcinomas. Am J Pathol 1990; 137:653-8.
103. Smith DM, Groff DE, Pokul RK, Bear JL, Delgado G. Determination of cellular oncogene rearrangement or amplification in ovarian adenocarcinomas. Am J Obstet Gynecol 1989; 161:911-5.
104. Yasue H, Takeda A, Ishibashi M. Amplification of the c-myc gene and the elevation of its transcripts in human ovarian tumor lines. Cell Struct Funct 1987; 12:121-5.
105. Sikora K, Evan G, Stewart J, Watson JV. Detection of the c-myc oncogene product in testicular cancer. Br J Cancer 1985; 52:171-6.
106. Fleming WH, Hamel A, MacDonald R, et al. Expression of the c-myc protooncogene in human prostatic carcinoma and benign prostatic hyperplasia. Cancer Res 1986; 46:1535-8.
107. Fox SB, Persad RA, Royds J, Kore RN, Silcocks PB, Collins CC. p53 and c-myc expression in stage A1 prostatic adenocarcinoma: useful prognostic determinants? J Urol 1993; 150:490-4.
108. Peehl DM. Oncogenes in prostate cancer. An update. [Review]. Cancer 1993; 71:1159-64.
109. Wolf DA, Kohlhuber F, Schulz P, Fittler F, Eick D. Transcriptional down-regulation of c-myc in human prostate carcinoma cells by the synthetic androgen mibolerone. Br J Cancer 1992; 65:376-82.
110. Buttyan R, Sawczuk IS, Benson MC, Siegal JD, Olsson CA. Enhanced expression of the c-myc protooncogene in high-grade human prostate cancers. Prostate 1987; 11:327-37.
111. Davies P, Eaton CL, France TD, Phillips ME. Growth factor receptors and oncogene expression in prostate cells. [Review]. Am J Clin Oncol 1988; 11:S1-7.
112. Thompson TC, Southgate J, Kitchener G, Land H. Multistage carcinogenesis induced by ras and myc oncogenes in a reconstituted organ. Cell 1989; 56:917-30.
113. Thompson TC, Park SH, Timme TL, et al. Loss of p53 function leads to metastasis in ras+myc-initiated mouse prostate cancer. Oncogene 1995; 10:869-79.

114. Adams JM, Cory S. Transgenic models of tumor development. [Review]. Science 1991; 254:1161-7.
115. Adams JM, Cory S. Transgenic models for haemopoietic malignancies. [Review]. Biochim Biophys Acta 1991; 1072:9-31.
116. Morgenbesser SD, DePinho RA. Use of transgenic mice to study myc family gene function in normal mammalian development and in cancer. [Review]. Semin Cancer Biol 1994; 5:21-36.
117. Adams JM, Harris AW, Pinkert CA, et al. The c-myc oncogene driven by immunoglobulin enhancers induces lymphoid malignancy in transgenic mice. Nature 1985; 318:533-8.
118. Alexander WS, Schrader JW, Adams JM. Expression of the c-myc oncogene under control of an immunoglobulin enhancer in Eμ-myc transgenic mice. Mol Cell Biol 1987; 7:1436-44.
119. Harris AW, Pinkert CA, Crawford M, Langdon WY, Brinster RL, Adams JM. The Eμ-myc transgenic mouse. A model for high-incidence spontaneous lymphoma and leukemia of early B cells. J Exp Med 1988; 167:353-71.
120. Schmidt EV, Pattengale PK, Weir L, Leder P. Transgenic mice bearing the human c-myc gene activated by an immunoglobulin enhancer: a pre-B-cell lymphoma model. Proc Natl Acad Sci USA 1988; 85:6047-51.
121. Langdon WY, Harris AW, Cory S, Adams JM. The c-myc oncogene perturbs B lymphocyte development in E-mu-myc transgenic mice. Cell 1986; 47:11-8.
122. Sinn E, Muller W, Pattengale P, Tepler I, Wallace R, Leder P. Coexpression of MMTV/v-Ha-ras and MMTV/c-myc genes in transgenic mice: synergistic action of oncogenes in vivo. Cell 1987; 49:465-75.
123. Haupt Y, Alexander WS, Barri G, Klinken SP, Adams JM. Novel zinc finger gene implicated as myc collaborator by retrovirally accelerated lymphomagenesis in E mu-myc transgenic mice [see comments]. Cell 1991; 65:753-63.
124. Lovec H, Grzeschiczek A, Kowalski MB, Moroy T. Cyclin D1/bcl-1 cooperates with myc genes in the generation of B-cell lymphoma in transgenic mice. EMBO J 1994; 13:3487-95.
125. Benvenisty N, Ornitz DM, Bennett GL, et al. Brain tumours and lymphomas in transgenic mice that carry HTLV-I LTR/c-myc and Ig/tax genes. Oncogene 1992; 7:2399-405.
126. van Lohuizen M, Verbeek S, Krimpenfort P, et al. Predisposition to lymphomagenesis in pim-1 transgenic mice: cooperation with c-myc and N-myc in murine leukemia virus-induced tumors. Cell 1989; 56:673-82.
127. Seldin DC, Leder P. Casein kinase II alpha transgene-induced murine lymphoma: relation to theileriosis in cattle [see comments]. Science 1995; 267:894-7.
128. Bodrug SE, Warner BJ, Bath ML, Lindeman GJ, Harris AW, Adams JM. Cyclin D1 transgene impedes lymphocyte maturation and col-

laborates in lymphomagenesis with the myc gene. EMBO J 1994; 13:2124-30.
129. Haupt Y, Bath ML, Harris AW, Adams JM. bmi-1 transgene induces lymphomas and collaborates with myc in tumorigenesis. Oncogene 1993; 8:3161-4.
130. Rosenbaum H, Harris AW, Bath ML, et al. An E mu-v-abl transgene elicits plasmacytomas in concert with an activated myc gene. EMBO J 1990; 9:897-905.
131. Schoenenberger CA, Zuk A, Groner B, Jones W, Andres AC. Induction of the endogenous whey acidic protein (Wap) gene and a Wap-myc hybrid gene in primary murine mammary organoids. Dev Biol 1990; 139:327-37.
132. Stewart TA, Pattengale PK, Leder P. Spontaneous mammary adenocarcinomas in transgenic mice that carry and express MTV/myc fusion genes. Cell 1984; 38:627-37.
133. Small JA, Blair DG, Showalter SD, Scangos GA. Analysis of a transgenic mouse containing simian virus 40 and v-myc sequences. Mol Cell Biol 1985; 5:642-8.
134. Leder A, Pattengale PK, Kuo A, Stewart TA, Leder P. Consequences of widespread deregulation of the c-myc gene in transgenic mice: multiple neoplasms and normal development. Cell 1986; 45:485-95.
135. Suda Y, Aizawa S, Hirai S, et al. Driven by the same Ig enhancer and SV40 T promoter ras induced lung adenomatous tumors, myc induced pre-B cell lymphomas and SV40 large T gene a variety of tumors in transgenic mice. EMBO J 1987; 6:4055-65.
136. Schoenenberger CA, Andres AC, Groner B, van der Valk M, LeMeur M, Gerlinger P. Targeted c-myc gene expression in mammary glands of transgenic mice induces mammary tumours with constitutive milk protein gene transcription. EMBO J 1988; 7:169-75.
137. Sandgren EP, Quaife CJ, Pinkert CA, Palmiter RD, Brinster RL. Oncogene-induced liver neoplasia in transgenic mice. Oncogene 1989; 4:715-24.
138. Jackson T, Allard MF, Sreenan CM, Doss LK, Bishop SP, Swain JL. The c-myc proto-oncogene regulates cardiac development in transgenic mice. Mol Cell Biol 1990; 10:3709-16.
139. Trudel M, D'Agati V, Costantini F. C-myc as an inducer of polycystic kidney disease in transgenic mice. Kidney Int 1991; 39:665-71.
140. Robbins RJ, Swain JL. C-myc protooncogene modulates cardiac hypertrophic growth in transgenic mice. Am J Physiol 1992; 262:H590-7.
141. Skoda RC, Tsai SF, Orkin SH, Leder P. Expression of c-MYC under the control of GATA-1 regulatory sequences causes erythroleukemia in transgenic mice. J Exp Med 1995; 181:1603-13.

CHAPTER 6

Properties of the c-Myc Protein

6.1 REGULATION OF c-Myc TRANSLATION

Biology is full of surprises, and the regulation of c-*myc* readily illustrates this point. The regulation of c-*myc* expression is an intricate network of commands that controls transcriptional initiation, elongation as well as mRNA stability. In addition to this hierarchy of controls, production of the c-Myc polypeptides is also regulated. The predicted size of the c-*myc* encoded polypeptide initiated at the canonical AUG translational start site is 439 amino acids.[1] The corresponding ATG is located at the 5' end of exon 2 (Fig. 4.1). Although the predicted molecular mass is 49.5 kDa, the observed size in sodium dodecyl sulfate polyacrylamide gel electrophoresis (SDS-PAGE) is about 62 kDa, which is largely due to the amino acid composition rather than to posttranslational modification.[2-4] Hence this protein is termed p62Myc. From the same c-*myc* mRNAs, an alternative form of c-Myc polypeptide is translationally initiated at a CUG, 14 codons upstream from the canonical AUG (Fig. 6.1).[5] This alternative form is termed p64Myc. These two forms of c-Myc appear as a doublet of polypeptides on SDS-PAGE.

Although the significance of p64Myc was not clear when it was discovered, Hann et al noted that many of the c-*myc* translocations in Burkitt's lymphomas resulted in the loss of this alternative form.[5] As such, only the predicted p62Myc is expressed in these lymphomas. If c-Myc contributes positively to Burkitt's lymphoma cell growth, these findings suggest that p62Myc contributes positively while p64Myc contributes negatively to cellular proliferation. Recently, Hann and coworkers pursued this idea and demonstrated that p64Myc is preferentially expressed in cells that cease to grow after achieving confluency or if starved of methionine.[6]

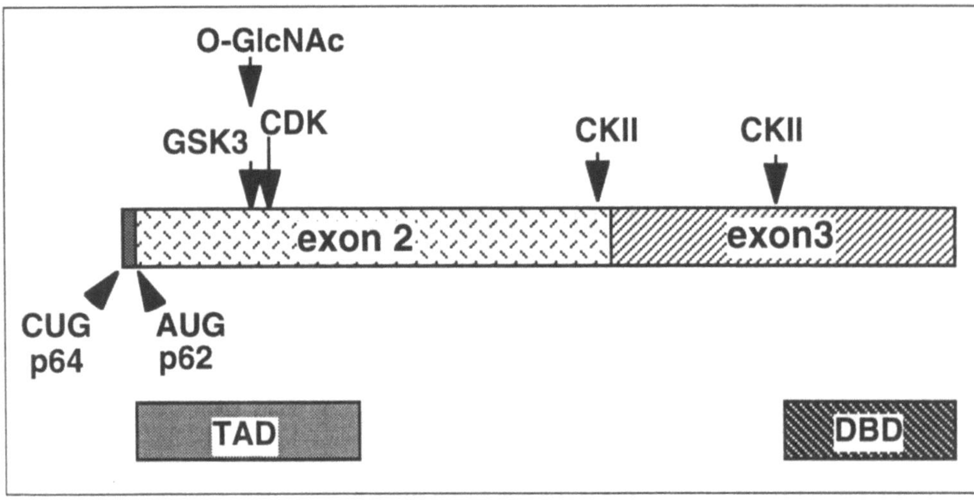

Fig. 6.1. Phosphorylation and glycosylation sites of c-Myc. p62 and p64 are two alternatively translated proteins that differ in size by 14 amino acids. The phosphorylation sites of c-Myc are indicated for those found in vivo. The putative kinases are indicated over the site of phosphorylation: GSK3, cyclin dependent kinase (CDK) and casein kinase II (CKII). An O-linked glycosylation site of c-Myc (O-GlcNAc) has been mapped to threonine-58. TAD = transactivation domain; DBD = DNA binding domain.

The biochemical and transforming properties of p62Myc and p64Myc are quite similar. The half-lives of these proteins vary from 15 to 60 minutes.[7,8] The ability of these proteins, as chimeric proteins with the yeast derived GAL4 protein or as intact proteins, to transactivate reporter plasmid constructs driven by GAL4 binding sites or the E-box, CACGTG are similar (see Chapters 7 and 8).[9] The abilities of these forms of c-Myc to transform Rat 1 cells in cooperation with Bcr-Abl appear equivalent.[10] Recent studies by Hann and coworkers, however, suggest that the transcriptional properties of p62Myc and p64Myc are distinguishable with a novel target DNA binding site (Chapters 7 and 8).[11] If these new observations are substantiated, then translational control of c-Myc is yet another mechanism to regulate the expression and function of c-*myc*.

A shorter form of c-Myc (45 kDa) is frequently detected in immunoprecipitates. Although this short c-Myc may be a result of proteolytic degradation, it appears that short c-Myc arises from an internal translational initiation site about 100 amino acids into the amino-terminal end. While this notion remains to be rigorously substantiated, the implication of such an amino-terminally truncated c-Myc is intriguing. Since the activation of c-Myc has been localized to the first 143 amino acids, a shortened c-Myc could behave as a naturally occurring dominant negative form of c-Myc.

6.2 c-Myc PROTEIN LEVELS

Due to its short half-life of about 20 minutes, c-Myc protein levels can rapidly change with cellular proliferative states. However, the levels of c-Myc do not appear to change with the different phases of the cell cycle.[7,8,12] The rate of degradation of c-Myc appears to be constant throughout the cell cycle except during mitosis when c-Myc is most stable. c-Myc protein level is low in quiescent swiss 3T3 and MRC-5 fibroblasts (1000-3000 molecules/cell), and its level is increased by about 2-fold (3000-6000 molecules/cell) with serum stimulation.[13] Similar changes in c-Myc levels are observed in resting and stimulated thymocytes. Certain malignant cell lines, such as HeLa cervical carcinoma or Raji Burkitt lymphoma cells have as many as 100,000 molecules of c-Myc per cell.[14] c-Myc is known to associate with at least one other protein, termed Max, for its function. Thus, the levels of c-Myc alone may not be sufficient to determine cellular fate.

In contrast to the elevation of c-Myc levels in cells stimulated to proliferate, various cell lines that are induced to differentiate display dramatic decreases in c-Myc levels. For example, the promyelocytic leukemia cell HL-60, when induced to differentiate by dimethylsulfoxide (DMSO) into granulocytes, showed a change from about 100,000 molecules per cell to less than 10,000 molecules per cell.[14] Murine erythroleukemia (MEL) cells induced to differentiate with DMSO also display significant decreases in c-Myc level. Despite a prolongation of c-Myc half-life, there is a decline in c-Myc levels after DMSO exposure.[15] These findings are somewhat different than results from another study in which c-Myc protein level did not decrease with MEL cell differentiation.[16] The half-life of c-Myc, however, was found to increase while c-*myc* mRNA level decreased with differentiation.

The short half-life of c-Myc protein has been attributed to its multiple PEST (proline-glutamate-serine-threonine) sequence motifs found commonly in short-lived proteins.[17] Although the exact cellular mechanism mediating PEST protein degradation is not yet clear, ubiquitination has been implicated as a possible mediator of this protein degradation pathway. In fact, a study by Ciechanover et al suggests that c-Myc protein is ubiquitinated prior to its degradation.[18] Studies by Luscher and Eisenman, however, could not demonstrate a role for ubiquitination in c-Myc degradation.[12] It should be noted that ubiquitin-conjugated proteins are very labile, and can be technically difficult to detect. Some clues to factors affecting c-Myc protein stability have been gathered. The rapid degradation of c-Myc appears to be mediated in the nuclear compartment, since cytochalasin-mediated enucleated cells have prolonged c-Myc half-lives.[19] A candidate degradation machinery is the 19S multiprotease that has been found in the nucleus. The degradation of c-Myc appears to be sensitive to heavy metal depletion by o-phenanthroline and treatment with either N-ethylmaleimide or chloroquine. The studies of c-Myc degradation are far from complete. In fact, it would be instructive to identify the structural determinants required for the rapid degradation of c-Myc. Specifically, since coding sequence mutations are now found in translocated c-*myc* alleles (see Chapter 5), it would be intriguing to determine whether mutation of the putative c-Myc degradation signal may create a long-lived oncogenic Myc protein.

6.3 POSTTRANSLATIONAL MODIFICATION: PHOSPHORYLATION AND GLYCOSYLATION

The c-Myc protein was rapidly recognized as a phosphoprotein since it is readily metabolically labelled with radioactive orthophosphate.[2,20-22]

The functional significance of c-Myc phosphorylation, however, remains incompletely understood. The overall phosphorylation state of c-Myc remains unaltered throughout the cell cycle.[8] It appears that c-Myc is phosphorylated by a number of kinases including casein kinase II (CKII),[23-25] MAP kinase,[26-29] DNA-dependent protein kinase,[26,30] and possibly cAMP-dependent protein kinase (PKA) and cdc2 (Fig. 6.1).[26] The sites of CKII phosphorylations have been mapped to amino acids 250 and 350;[25] however, the biological consequence of CKII-mediated phosphorylation of c-Myc remains unknown.[24] It is intriguing, nevertheless, that the crossing of transgenic *myc* mice with CKII transgenic animals resulted in accelerated lymphomagenesis.[31] CKII phosphorylation of c-Myc does not appear to affect its ability to bind specific DNA sequences as a heterodimer with Max.[32,33] The sites of DNA-dependent protein kinase or cAMP-dependent protein kinase phosphorylation on c-Myc are not known. Moreover, c-Myc phosphopeptides generated by DNA-dependent protein kinase or cAMP-dependent kinase in vitro do not resemble in vivo labelled phosphopeptides suggesting that these kinases may not play a physiological role in regulating c-Myc.[26]

MAP kinase has been shown to phosphorylate within the c-Myc transcriptional activation domain (amino acids 58 and 62) with consequent attenuation of transcriptional potential.[27,28,34] Whether MAP kinase phosphorylates Myc in vivo remains unestablished.[26] Recent studies with Myc proteins mutated at Thr-58 or Ser-62 did not confirm the alteration in transactivation by Myc.[26,35] Two studies, however, suggest that mutation of Thr-58 increased the transforming activity of c-Myc.[35,36] Ser-62 mutations increased Myc transforming activity in the Rat 1a anchorage-independent growth assay but appeared to decrease Myc co-transforming activity with an activated *ras* gene.[35,36] Scrutiny of the N-terminal phosphorylation sites revealed that phosphorylation of Thr-58 follows prior phosphorylation of Ser-62.[26] The dependence of site-specific phosphorylation on the cell cycle is not yet clear. In one study the phosphorylation of Ser-62 appears attenuated in mitosis.[26] In contrast another study suggests that the phosphorylation of Ser-62 appears to be highest in the G_2-M phase of the cell cycle.[34] These differences are unlikely to be the result of the use of different cell lines. Rather, the techniques of cell synchronization and selection of cell populations may alter the phosphorylation state of c-Myc dramatically. Recent studies with mutant *myc* alleles derived from lymphomas have shed additional light on the significance of Myc phosphorylation (see chapter 7).[37]

A currently held hypothesis is that O-linked glycosylation of nuclear proteins is as dynamic as phosphorylation. Although the exact functional consequence of O-linked glycosylation is not understood, a number of transcriptional factors are glycosylated as is RNA polymerase II. Similarly, c-Myc protein appears to be O-linked glycosylated at its amino-terminal end.[38] In fact, purified c-Myc protein over-produced in insect cells, has been subjected to tryptic proteolysis, labelled with gal-transferase, and analyzed by manual Edman degradation, mass spectrometry and amino acid sequencing. Intriguingly, a major glycosylation site has been mapped to threonine-58, a known phosphorylation site and a mutational hotspot in lymphomas.[39] The stoichiometry of modified residues in c-Myc and the function of O-linked glycosylation of c-Myc is currently being investigated.

6.4 NUCLEAR LOCALIZATION AND ASSOCIATION WITH THE NUCLEAR MATRIX

Early immunological studies indicate that c-Myc (or v-Myc) is a nuclear protein that is detected in the nucleoplasm excluding the nucleoli.[21,40-44] One morphologic study showed an overlap of c-Myc specific immunofluorescence and small ribonuclear proteins (sRNPs) suggesting the possible role of c-Myc in RNA processing.[45] This observation, however, could not be further confirmed by biochemical studies to demonstrate an association of c-Myc with isolated sRNPs. Recently, some evidence for the subcellular redistribution of c-Myc corresponding to cell proliferation states has been reported. The data suggest that in non-dividing cells immunologically reactive Myc protein is found in the cytoplasmic compartment.[46] In addition, c-Myc protein stored in *Xenopus* oocytes appears to be virtually all cytoplasmic. Subsequent to fertilization c-Myc protein is translocated into the nuclear compartment.[47,48] Another study, however, did not confirm this observation.[49] This phenomenon, particularly in mammalian cells, needs to be further verified, and the molecular basis for subcellular redistribution of c-Myc requires additional careful studies.

The c-Myc protein has also been shown to be associated with the nuclear matrix, which is the residuum that remains after detergent and salt extraction and nuclease treatment of isolated nuclei.[50] Components of the DNA replication and transcriptional machinery have been found associated with the nuclear matrix. Such findings suggest the existence of an organized semi-solid matrix for biosynthetic machineries. With standard preparations of nuclear matrices, c-Myc association with the nuclear matrix appears to be dependent on cupric ions.[51] Other studies suggest, however, that the association of c-Myc with the nuclear matrix may be an artifactual consequence of thermal denaturation.[52] That is, matrices prepared in low temperatures are devoid of c-Myc protein. The latter observation does not rule out, however, that c-Myc associates with some nuclear matrix structures intracellularly.

c-Myc overexpression has been shown to result in the association of c-Myc proteins with heat shock protein 70 (HSP 70) in the nucleus.[53,54] The amino-terminal domain of the c-Myc protein is required for association with HSP 70, although the functional significance of this association is as yet unknown. However, since overexpressed c-Myc accumulated in large amorphous bodies with HSP 70, it is likely that this association reflects how cells contend with highly overexpressed proteins. It is notable that overexpression of Myc proteins alters the ultrastructure of nuclear chromatin.[55]

REFERENCES

1. Watt R, Stanton LW, Marcu KB, Gallo RC, Croce CM, Rovera G. Nucleotide sequence of cloned cDNA of human c-myc oncogene. Nature 1983; 303:725-8.
2. Ramsay G, Evan GI, Bishop JM. The protein encoded by the human proto-oncogene c-myc. Proc Natl Acad Sci USA 1984; 81:7742-6.
3. Alitalo K, Ramsay G, Bishop JM, Pfeifer SO, Colby WW, Levinson AD. Identification of nuclear proteins encoded by viral and cellular myc oncogenes. Nature 1983; 306:274-7.
4. Giallongo A, Appella E, Ricciardi R, Rovera G, Croce CM. Identification of the c-myc oncogene product in normal and malignant B cells. Science 1983; 222:430-2.
5. Hann SR, King MW, Bentley DL, Anderson CW, Eisenman RN. A non-AUG translational initiation in c-myc exon 1 generates an N-terminally distinct protein whose synthesis is disrupted in Burkitt's lymphomas. Cell 1988; 52:185-95.
6. Hann SR, Sloan-Brown K, Spotts GD. Translational activation of the non-AUG-initiated c-myc 1 protein at high cell densities due to methionine deprivation. Genes Dev 1992; 6:1229-40.
7. Rabbitts PH, Watson JV, Lamond A, et al. Metabolism of c-myc gene products: c-myc mRNA and protein expression in the cell cycle. EMBO J 1985; 4:2009-15.
8. Hann SR, Thompson CB, Eisenman RN. c-myc oncogene protein synthesis is independent of the cell cycle in human and avian cells. Nature 1985; 314:366-9.
9. Kato GJ, Barrett J, Villa-Garcia M, Dang CV. An amino-terminal c-myc domain required for neoplastic transformation activates transcription. Mol Cell Biol 1990; 10:5914-20.
10. Blackwood EM, Lugo TG, Kretzner L, et al. Functional analysis of the AUG- and CUG-initiated forms of the c-Myc protein. Mol Biol Cell 1994; 5:597-609.
11. Hann SR, Dixit M, Sears RC, Sealy L. The alternatively initiated c-Myc proteins differentially regulate transcription through a noncanonical DNA-binding site. Genes Dev 1994; 8:2441-52.
12. Luscher B, Eisenman RN. c-myc and c-myb protein degradation: effect of metabolic inhibitors and heat shock. Mol Cell Biol 1988; 8:2504-12.
13. Waters CM, Littlewood TD, Hancock DC, Moore JP, Evan GI. c-myc protein expression in untransformed fibroblasts. Oncogene 1991; 6:797-805.
14. Moore JP, Hancock DC, Littlewood TD, Evan GI. A sensitive and quantitative enzyme-linked immunosorbence assay for the c-myc and N-myc oncoproteins. Oncogene Res 1987; 2:65-80.
15. Spotts GD, Hann SR. Enhanced translation and increased turnover of c-myc proteins occur during differentiation of murine erythroleukemia cells. Mol Cell Biol 1990; 10:3952-64.

16. Wingrove TG, Watt R, Keng P, Macara IG. Stabilization of myc proto-oncogene proteins during Friend murine erythroleukemia cell differentiation. J Biol Chem 1988; 263:8918-24.
17. Rogers S, Wells R, Rechsteiner M. Amino acid sequences common to rapidly degraded proteins: the PEST hypothesis. Science 1986; 234:364-8.
18. Ciechanover A, DiGiuseppe JA, Bercovich B, et al. Degradation of nuclear oncoproteins by the ubiquitin system in vitro. Proc Natl Acad Sci USA 1991; 88:139-43.
19. Luscher B, Eisenman RN. Proteins encoded by the c-myc oncogene: analysis of c-myc protein degradation. Princess Takamatsu Symp 1986; 17:291-301.
20. Hann SR, Eisenman RN. Proteins encoded by the human c-myc oncogene: differential expression in neoplastic cells. Mol Cell Biol 1984; 4:2486-97.
21. Persson H, Gray HE, Godeau F, Braunhut S, Bellve AR. Multiple growth-associated nuclear proteins immunoprecipitated by antisera raised against human c-myc peptide antigens. Mol Cell Biol 1986; 6:942-9.
22. Ramsay G, Hayman MJ, Bister K. Phosphorylation of specific sites in the gag-myc polyproteins encoded by MC29-type viruses correlates with their transforming ability. EMBO J 1982; 1:1111-6.
23. Hagiwara T, Nakaya K, Nakamura Y, Nakajima H, Nishimura S, Taya Y. Specific phosphorylation of the acidic central region of the N-myc protein by casein kinase II. Eur J Biochem 1992; 209:945-50.
24. Bousset K, Oelgeschlager MHH, Henriksson M, et al. Regulation of transcription factors c-Myc, Max, and c-Myb by casein kinase II. Cell Molec Biol Res 1995; 40:501-511.
25. Luscher B, Kuenzel EA, Krebs EG, Eisenman RN. Myc oncoproteins are phosphorylated by casein kinase II. EMBO J 1989; 8:1111-9.
26. Lutterbach B, Hann SR. Hierarchical phosphorylation at n-terminal transformation-sensitive sites in c-myc protein is regulated by mitogens and in mitosis. Mol Cell Biol 1994; 14:5510-22.
27. Seth A, Alvarez E, Gupta S, Davis RJ. A phosphorylation site located in the NH2-terminal domain of c-Myc increases transactivation of gene expression. J Biol Chem 1991; 266:23521-4.
28. Gupta S, Seth A, Davis R. Transactivation of gene expression by Myc is inhibited by mutation at the phosphorylation sites Thr-58 and Ser-62. Proc Natl Acad Sci, USA 1993; 90:3216-20.
29. Jackson SP. Regulating transcription factor activity by phosphorylation. TICB 1992; 2:104-8.
30. Iijima S, Teraoka H, Date T, Tsukada K. DNA-activated protein kinase in Raji Burkitt's lymphoma cells. Phosphorylation of c-Myc oncoprotein. Eur J Biochem 1992; 206:595-603.

31. Seldin DC, Leder P. Casein kinase II alpha transgene-induced murine lymphoma: relation to theileriosis in cattle [see comments]. Science 1995; 267:894-7.
32. Berberich SJ, Cole MD. Casein kinase II inhibits the DNA-binding activity of Max homodimers but not Myc/Max heterodimers. Genes Dev. 1992; 6:166-76.
33. Bousset K, Henriksson M, Luscherfirzlaff JM, Litchfield DW, Luscher B. Identification of casein kinase ii phosphorylation sites in max - effects on dna-binding kinetics of max homo- and myc/max heterodimers. Oncogene 1993; 8:3211-20.
34. Seth A, Gupta S, Davis RJ. Cell cycle regulation of the c-Myc transcriptional activation domain. Mol Cell Biol 1993; 13:4125-36.
35. Henriksson M, Bakardjiev A, Klein G, Luscher B. Phosphorylation sites mapping in the N-terminal domain of c-*myc* modulate its transforming potential. Oncogene 1993; 8:3199-209.
36. Pulverer BJ, Fisher C, Vousden K, Littlewood T, Evan G, Woodgett JR. Site-specific modulation of c-myc cotransformation by residues phosphorylated in vivo. Oncogene 1994; 9:59-70.
37. Hoang AT, Lutterbach B, Lewis BC, et al. A link between increased transforming activity of lymphoma-derived MYC mutant alleles, their defective regulation by p107, and altered phosphorylation of the c-Myc transactivation domain. Mol Cell Biol 1995; 15:4031-42.
38. Chou TY, Dang CV, Hart GW. Glycosylation of the c-Myc transactivation domain. Proc Natl Acad Sci USA 1995; 92:4417-21.
39. Chou TY, Hart GW, Dang CV. c-Myc is glycosylated at threonine-58, a known phosphorylation site and a mutational hot spot in lymphomas. J Biol Chem 1995; 270.
40. Hann SR, Abrams HD, Rohrschneider LR, Eisenman RN. Proteins encoded by v-myc and c-myc oncogenes: identification and localization in acute leukemia virus transformants and bursal lymphoma cell lines. Cell 1983; 34:789-98.
41. Abrams HD, Rohrschneider LR, Eisenman RN. Nuclear location of the putative transforming protein of avian myelocytomatosis virus. Cell 1982; 29:427-39.
42. Bunte T, Greiser-Wilke I, Donner P, Moelling K. Association of gag-myc proteins from avian myelocytomatosis virus wild-type and mutants with chromatin. EMBO J 1982; 1:919-27.
43. Bunte T, Greiser-Wilke I, Moelling K. The transforming protein of the MC29-related virus CMII is a nuclear DNA-binding protein whereas MH2 codes for a cytoplasmic RNA-DNA binding polyprotein. EMBO J 1983; 2:1087-92.
44. Persson H, Leder P. Nuclear localization and DNA binding properties of a protein expressed by human c-myc oncogene. Science 1984; 225:718-21.

45. Spector DL, Watt RA, Sullivan NF. The v- and c-myc oncogene proteins colocalize in situ with small nuclear ribonucleoprotein particles. Oncogene 1987; 1:5-12.
46. Craig RW, Buchan HL, Civin CI, Kastan MB. Altered cytoplasmic/nuclear distribution of the c-myc protein in differentiating ML-1 human myeloid leukemia cells. Cell Growth Differ 1993; 4:349-57.
47. Gusse M, Ghysdael J, Evan G, Soussi T, Mechali M. Translocation of a store of maternal cytoplasmic c-myc protein into nuclei during early development. Mol Cell Biol 1989; 9:5395-403.
48. Vriz S, Lemaitre JM, Leibovici M, Thierry N, Mechali M. Comparative analysis of the intracellular localization of c-Myc, c-Fos, and replicative proteins during cell cycle progression. Mol Cell Biol 1992; 12:3548-55.
49. King MW, Roberts JM, Eisenman RN. Expression of the c-myc proto-oncogene during development of Xenopus laevis. Mol Cell Biol 1986; 6:4499-508.
50. Eisenman RN, Tachibana CY, Abrams HD, Hann SR. V-myc- and c-myc-encoded proteins are associated with the nuclear matrix. Mol Cell Biol 1985; 5:114-26.
51. Van Straaten JP, Rabbitts TH. The c-myc protein is associated with the nuclear matrix through specific metal interaction. Oncogene Res 1987; 1:221-8.
52. Evan GI, Hancock DC. Studies on the interaction of the human c-myc protein with cell nuclei: p62c-myc as a member of a discrete subset of nuclear proteins. Cell 1985; 43:253-61.
53. Koskinen PJ, Sistonen L, Evan G, Morimoto R, Alitalo K. Nuclear colocalization of cellular and viral myc proteins with HSP70 in myc-overexpressing cells. J Virol 1991; 65:842-51.
54. Henriksson M, Classon M, Axelson H, Klein G, Thyberg J. Nuclear colocalization of c-myc protein and hsp70 in cells transfected with human wild-type and mutant c-myc genes. Exp Cell Res 1992; 203:383-94.
55. Henriksson M, Classon M, Ingvarsson S, et al. Elevated expression of c-myc and N-myc produces distinct changes in nuclear fine structure and chromatin organization. Oncogene 1988; 3:587-93.

CHAPTER 7

Functional Domains of the c-Myc Protein

7.1 NEOPLASTIC TRANSFORMATION

The ability of c-*myc* to transform cells in transgenic animals and its deregulated expression in human cancers provide evidence that c-*myc* plays a central role in neoplastic transformation. Its transforming activity can be assayed in tissue cultured cells as well. c-*myc* requires cooperation with an activated *ras* gene to transform primary rat embryo cells (RECs) in a focus formation assay[1,2] (Fig. 1.12), whereas c-*myc* alone is able to transform a derivative rat embryo fibroblast cell line called Rat 1a.[3,4] In the avian system, v-Myc transforms macrophages and fibroblasts.[5,6]

The regions of c-Myc necessary for transforming RECs or rat 1a cells were determined through the use of in-frame insertion and deletion mutants of the normal human c-*myc* coding domain that was expressed from a retroviral promoter-enhancer (Fig. 7.1).[4] In the REC cotransformation assay, the amino-terminal 143 amino acids and carboxy-terminal 89 amino acids were required for activity. Small deletions within the first 90 amino acids were well-tolerated, whereas removal of amino acids 106-143 completely abolished activity. An insertional mutation at codon 106 was detrimental indicating that it is essential for transformation. A separate study specifically targeted c-Myc sequences conserved among *myc* family members, termed Myc boxes I and II, for mutagenesis.[7] In this study, c-Myc box II (amino acids 122-144) mutants were most dramatically affected, whereas Myc box I (amino acids 55-65) mutants still retained activity.

The regions required for transformation of rat 1a cells are identical to those involved in transforming RECs with the exception of the middle portion (residues 143 to 262) of c-Myc, which

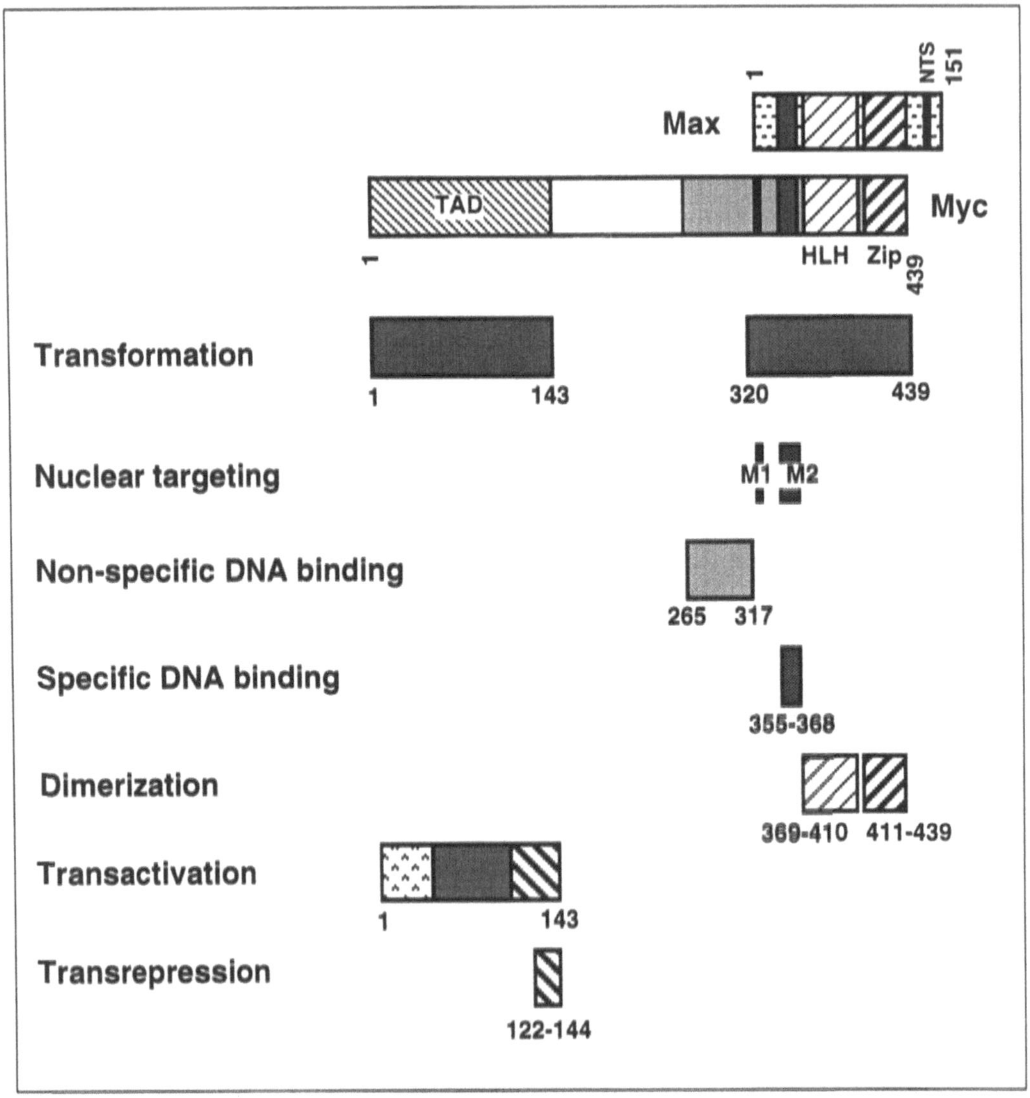

Fig. 7.1. Functional domains of c-Myc. The bars at the top summarize the domains of Myc and Max. The numbers correspond to amino acid residues. Transformation of rat embryo cells in cooperation with an activated ras gene requires the integrity of two regions of c-Myc. The nuclear targeting sequences, M1 and M2, are indicated. The nonspecific DNA binding domain was mapped to amino acids 265 to 317. The specific DNA recognition constitutes the basic region that is conserved in helix-loop-helix (HLH) proteins. The dimerization domain is composed of an HLH domain that extends into a leucine zipper (Zip). The tripartite transactivation domain is depicted. Deletion of amino acids 122 to 144 has been shown to abrogate the ability of c-Myc to transrepress initiator driven transcription.

contains an acidic (polyglutamate) region.[4] Deletion of this region does not abolish REC transforming activity, but its removal dramatically decreases Rat 1a transformation. The homologous region of v-Myc also has cell-type specific effect. The removal of the acidic region abrogates v-Myc transforming activity in hematopoietic cells but not in fibroblasts.[8] Furthermore, these mutants were able to transform quail macrophages but not chicken macrophages.[9] These observations suggest that certain cell types or assays have a less stringent requirement for full Myc activity. Although the function of the central acidic region of Myc is unknown, its biological manifestations are clear.

7.2 CELLULAR DIFFERENTIATION

In several cell culture models deregulated expression of c-Myc blocks cell differentiation. Although cellular transformation is thought to require a block in cell differentiation programs, the existence of malignancies with differentiated phenotypes is well recognized by anatomic pathologists. Thus, the fact that c-Myc transforms cells through specific domains does not necessarily imply that the same regions inhibit cell differentiation. Several investigators have used c-Myc mutants in differentiation models to determine the domains of c-Myc required for blocking differentiation.

As described in chapter 1, the 3T3L1 preadipocytic fibroblasts were induced to differentiate into adipocytes, as measured by the production of glycerophosphate dehydrogenase, under defined culture conditions.[10] The deregulated expression of c-*myc* blocks differentiation in a manner that is dependent on the amino-terminal 143 amino acids and the carboxy-terminal 89 amino acids of c-Myc. The central acidic region is also required for c-Myc activity.

In separate studies, murine erythroleukemia cells expressing wild-type and mutant c-*myc* alleles were induced to differentiate.[11-13] Erythrocytic differentiation was determined by the production of hemoglobin. Early in the induction of murine erythroleukemia (MEL) cell differentiation, c-*myc* mRNA levels dramatically decrease.[14,15] The deregulated expression of a transfected c-*myc* gene inhibited the commitment and differentiation of MEL cells. The c-*myc* mutant genes that lack sequences overlapping the putative transcriptional activation domain, the helix-loop-helix motif or the leucine zipper motif failed to inhibit N,N'-hexamethylene-bis-acetamide-induced differentiation of MEL cells.[12] Two partially deleted c-*myc* genes, one lacking a short NH_2-terminal region and the other lacking 118 amino acids at the center of the coding region, transform RECs but do not block MEL differentiation.

In yet another study, human c-*myc* mutants transfected into MEL cells under an inducible promoter revealed that the c-Myc domains necessary for commitment and differentiation are not co-localized; almost the entire c-Myc is required for inhibition of commitment, whereas the amino-terminal and carboxy-terminal regions that are essential for co-transforming activity with *ras* are required for inhibition of differentiation.[13] Mutants that delete domains for c-Myc dimerization motifs enhanced differentiation suggesting that c-Myc may regulate commitment and differentiation by interacting with proteins through different domains or through a dominant negative effect. It has been demonstrated that the amino-terminal transactivation domain of c-Myc, when tethered to GAL4, is able to inhibit c-Myc co-transforming activity.[16,17] It is therefore possible that mutants with deletion of the dimerization motifs may inhibit growth and thereby enhance differentiation through a dominant negative effect.

7.3 DNA REPLICATION

c-Myc has been hypothesized to participate in DNA replication. In an early study, microinjection of c-Myc protein made cells competent to undergo DNA replication and complete the cell cycle.[18] The ability of normal c-Myc and Myc deletion mutants to induce serum starved, untransformed NIH3T3 cells to enter S phase was analyzed by microinjecting these quiescent cells with wild type and mutant human c-*myc* plasmids.[19] Full length c-Myc is able to induce S phase entry, but c-Myc mutants that delete amino acids 7-91, 41-53, 56-103, 106-143, 265-317 and 414-433 are completely inactive. Myc overexpressing cells entered S phase with the same kinetics as control cells when both were stimulated with 20% fetal calf serum (FCS). c-Myc overexpression, however, did increase the percentage of cells entering S phase upon serum stimulation. Myc mutants that have a deletion amino acids 106-143 or 414-433 exerted a dominant negative effect on S phase entry both in quiescent cells stimulated with 2% FCS and in unsynchronized, cycling cells. These studies as well as earlier studies firmly established the ability of c-Myc overexpression to enforce S-phase entry; however, whether this involves a direct interaction of c-Myc with the replication machinery is unknown.

Several studies have provided evidence that c-Myc directly participates in DNA replication or regulates the replication machinery.[20] DNA replication in isolated nuclei was observed to be augmented by the addition of purified c-Myc protein.[21] In addition, an anti-Myc antibody was shown to inhibit DNA synthesis measured in permeabilized isolated nuclei. The evidence provided by this study appeared incontrovertible; however, it was subsequently discovered that the anti-Myc antibody was contaminated with an unknown inhibitor of DNA synthesis.[22,23] Furthermore, purified c-Myc protein appeared to be significantly contaminated with bacterial DNA polymerase that could result in an apparent stimulation of DNA synthesis in these assays (Dang, unpublished observation).

Observations made by Iguchi-Ariga and coworkers suggest that the c-Myc protein binds directly to mammalian origins of replication.[24-27] One such putative origin lies about 2 kb upstream of the *c-myc* coding sequence. The putative Myc binding site contains the sequence, TCTCTTA, that was later found to compete with the E-box sequence CACGTG in gel mobility shift assays. This sequence appears to be transactivated by c-Myc as well. The binding of c-Myc to this sequence, however, appears weak as it has been difficult to reproduce. Nevertheless, the ability of the immediate 2.4 kb 5' sequence of *c-myc* to serve as a replication origin has been independently verified, although the minimal origin has not been mapped.[28]

The participation of c-Myc in DNA replication has also been suggested by observations that c-Myc is able to replace SV40 large T antigen in the replication of SV40 origin-containing plasmids.[25,29] The replicating activity of SV40 origin-containing plasmid was tested in human HL-60 and Raji cells and in monkey Cos1 cells. Even in the absence of SV40 T antigen, plasmids possessing SV40 ori sequences could replicate in human cells which are expressing high levels of *c-myc*.[25] Intriguingly, replication of the SV40 ori-containing plasmids in HL-60 cells was inhibited by anti-c-Myc antibody introduced into the cells. In a separate study, the SV40-based vector, pSVEpR4, was found to replicate poorly in the human BJAB lymphoma line and in most human cells.[29] This plasmid, however, replicates well in Burkitt's lymphoma lines with high c-*myc* expression. Cotransfection of the BJAB cells with a c-*myc*-expressing construct increased the replication of pSVEpR4 tenfold indicating that overexpression of the c-*myc* gene product allows the replication of SV40 in human lymphoma cells. A series of c-*myc* deletion mutants was used to investigate which c-Myc regions are important in stimulating SV40 replication.[30] The ability of c-Myc to promote SV40 replication was significantly reduced by deletions in the transactivation domain as well as in the basic region or leucine zipper motif. Other Myc members, L-Myc and N-Myc, also have replication stimulatory activity.[31] In sum, these observations suggest that the Myc proteins can substitute for SV40 T antigen in SV40 DNA replication.

A putative origin of replication has been isolated and found to contain the E-box sequence CGTCACGTGATG.[32-34] This sequence was in Cot_0 DNA obtained at the onset of S phase in HL60 cells.[32] The isolated sequences, however, did not have autonomously replicating activity. Although c-Myc is not directly implicated, the presence of E-box sequences in putative origins of replication is not surprising, since transcriptional elements are components of origins of replications.[35-37] In fact, identification of the general transcription factor, TFII H, which contains helicase activity, provides a link between DNA replication and transcription.[38] The evidence for a direct participation of c-Myc in DNA replication is presently lacking; nonetheless as we learn more about the role of transcription in DNA replication a new function for c-Myc in S-phase may be more firmly established.

7.4 AUTOSUPPRESSION

Expression of the translocated c-*myc* allele in Burkitt's lymphomas is invariably associated with the suppression of the normal allele suggesting that c-*myc* expression may be regulated by a negative feedback loop. Negative feedback regulation of c-*myc* gene expression has been observed in some cell types.[39] In one study, the existence of negative c-*myc* autoregulation was observed when the rapid, dose dependent and reversible suppression of endogenous c-*myc* expression occurred in EBV-immortalized B lymphoblastoid cells transfected with an inducible c-*myc* gene.[40] The c-*myc* autoregulatory mechanism was found in all (11 of 11) non-tumorigenic cells tested, including normal and immortalized lymphocytes and fibroblasts. However, this mechanism was inoperative in all (10 of 10) tumor cell lines derived from a variety of tissues including those carrying normal and oncogenically activated c-*myc* genes. The mechanism by which c-Myc autoregulates expression is not defined, although the recent observation that c-Myc is able to repress transcription through initiator elements led to the hypothesis that autoregulation may involve initiator elements in the various c-*myc* promoters.[41] Formal evidence for the existence of this initiator-mediated feedback loop is pending.[42] Amino acids 106 to 143 (activation domain) and 354 to 433 (helix-loop-helix-zipper) of the human c-Myc protein were shown to be required for the c-Myc autosuppression and were found to exactly overlap with those necessary for c-*myc* to cooperate with *ras* oncogenes in the transformation of rat embryo fibroblasts.[43] It is noteworthy that deletion of amino acids 106-143, which is required for repression of initiator dependent transcription, also abrogates autosuppression.[41]

7.5 NUCLEAR LOCALIZATION

v-Myc and c-Myc were found to localize to the cell nucleus soon after the production of antibodies against Myc proteins.[44-47] Both cell fractionation studies as well as immunofluorescent microscopic studies identified the nuclear location of c-Myc in cells overexpressing Myc. More recently, mammalian and frog c-Myc have been found to be cytoplasmic under conditions of attenuated cellular growth as discussed in chapter 6. Additional studies to verify these observations will be required to determine whether compartmentation of Myc protein plays a major role in regulating its function.[48,49] Although the mechanism by which c-Myc is retained in the cytoplasm is unknown, the sequences responsible for c-Myc nuclear localization have been identified.

Prior to the study of the c-Myc nuclear localization signal, only a limited number of nuclear proteins were determined to bear sequences that direct them to the nucleus. The prototypic nuclear localization signal (PKRRRR) is that of the SV40 large T antigen.[50] Based on seminal studies of SV40 large T antigen and polyoma T antigen, sequences within c-Myc that can direct the protein to the nucleus were sought.[51] Deletion analysis indicated that removal of c-Myc amino acids 320 to 370 resulted in a predominantly cytoplasmic mutant protein, although nuclear staining was detected by immunofluorescent microscopy. To localize the nuclear targeting sequences, segments of the c-Myc protein were fused to the cytoplasmic protein pyruvate kinase to determine whether any of the c-Myc segments could "drag" the cytoplasmic marker into the nucleus. c-Myc amino acids 320 to 327 (PAAKRVKLD), termed M1, were able to target pyruvate kinase completely into the nucleus, whereas amino acids 350 to 364 (RQRRNELKRSP), termed M2, partially targeted pyruvate kinase to the nucleus (Fig. 7.1). Synthetic peptides bearing M1 or M2 were conjugated to serum albumin and injected into the cytoplasmic compartment of Vero cells to determine the kinetics of nuclear entry of these conjugates. M1 mediated rapid entry of the conjugated albumin into the nucleus within 5 minutes of microinjection. M2 caused only partial nuclear localization of the serum albumin conjugate even 2 hours after microinjection. These results suggested that M1 is the major nuclear targeting signal. M2 turned out to be the basic region of c-Myc that contacts specific DNA sequences directly. Whether the partial nuclear localization by M2 is due to some DNA binding activity of the chimeras or due to weak nuclear transport signal is unclear. Nevertheless, with more nuclear targeting signals identified in various nuclear proteins, it appears that nucleic acid binding domains frequently overlap with nuclear localization signals.

To determine whether nuclear localization is required for c-Myc function, deletion mutations directed at M1 or M2 were generated. Deletion of M1 resulted in a partially nuclear mutant that retained transforming activity. Deletion of M2, however, rendered the fully nuclear mutant inactive for transformation. This is not surprising in retrospect since M2 is the DNA-binding basic region of c-Myc. Thus, this early study indicates that nuclear localization is insufficient for transformation and that other domains within c-Myc are required. Recently, site-directed mutations in v-Myc that render the protein cytoplasmic did not abrogate its transforming activity.[52] The small amount of nuclear mutant v-Myc was indicated to be less than the amount of Max protein available in infected cells. It was not ruled out however that a small fraction of nuclear Myc might be sufficient for transformation or indeed that nuclear localization is not required for Myc transforming activity. While the suggestion that Myc could function in the cytoplasmic compartment is intriguing, it is not compatible with the findings that domains of c-Myc required for transcriptional modulatory activities are required for transformation.

7.6 NONSPECIFIC DNA BINDING

Since the c-Myc protein localized to the nuclear compartment, it was hypothesized that c-Myc is a DNA-binding protein.[5,44,53-57] The c-Myc protein has been known to bind DNA nonspecifically in vitro, however it was not known whether this property reflects the potential for c-Myc to bind specific DNA sequences or to transform cells. The ability of the c-Myc protein to bind DNA was first determined by chromatography on DNA cellulose.[55,58] Metabolically labelled extracts from cells that overexpressed v-Myc or c-Myc were chromatographed in a matrix composed of native calf thymus DNA immobilized on cellulose. Both v-Myc and c-Myc were found to associate with DNA relatively tightly.[56,57] This interaction could be disrupted with ionic strengths that exceeded 200 to 300 mM NaCl (or equivalent conditions). These observations suggest that Myc proteins bind DNA, however it was not clear as to the specificity of this interaction. Since most specific DNA binding proteins also associate with DNA nonspecifically, the DNA binding properties of Myc became an area of intense interest.

Early studies of v-Myc and its mutants suggest that its nonspecific DNA-binding properties correlate with transforming properties. v-Myc ($p110^{gag\text{-}myc}$) from the avian myelocytomatosis virus MC29 is a 110,000 kDa fusion of an N-terminal gag protein and Myc sequences that bind chromatin.[56] Three deletion mutants of MC29 were isolated and found to have a hematopoietic cell transforming potential that is 1% as efficient as wild-type v-Myc.[59] These mutants, however, transform fibroblasts as well as the wild-type virus. The corresponding mutant v-Myc proteins are $p100^{gag\text{-}myc}$, $p95^{gag\text{-}myc}$ and $p90^{gag\text{-}myc}$. All three proteins have deletions in the Myc sequences. These mutants had diminished DNA-binding affinity compared to wild-type MC29 $p110^{gag\text{-}myc}$ as assessed by native DNA-immunoaffinity chromatography.[59] These v-*myc* mutants were subsequently molecularly cloned and sequenced.[5] The deletions in $p100^{gag\text{-}myc}$, $p95^{gag\text{-}myc}$ and $p90^{gag\text{-}myc}$ overlap with a corresponding region of human c-Myc containing amino acids 265 to 317. All of these mutants retain the basic carboxy-terminal region. The later study, however, did not confirm the differences in DNA binding affinities among these mutants as reported earlier.[5]

The property of nonspecific DNA binding appears to involve a domain distinct from one that participates in specific DNA binding.

This issue is controversial in part due to different assays for DNA binding. To determine the nonspecific DNA binding domain of human c-Myc, the DNA binding properties of deletion mutant Myc proteins were compared with those of the wild-type protein produced in bacteria.[60] Amino acids 265 to 318 appear to be necessary for c-Myc to bind DNA nonspecifically as determined by southwestern blotting. A region of c-Myc that is sufficient for nonspecific DNA binding was further delineated by the demonstration that fusion of the non-DNA binding staphylococcal protein A to Myc amino acids 265-318 resulted in a DNA binding chimeric protein. This region of c-Myc overlaps with a region previously shown to be required for v-Myc to bind DNA. It is also required for the ability of c-*myc* to transform rat 1a cells but not for transformation of rat embryo cells in cooperation with activated ras.

Four Ser-Pro-X-X-like (X= other amino acids, commonly lysines) motif sequences, which have been shown to mediate nonspecific DNA binding by histone H1, are found within the Myc nonspecific DNA binding domain.[61,62] This observation suggests that the nonspecific DNA binding domain of c-Myc might facilitate localization of specific target DNA sites by reducing the dimensionality of diffusion through binding to nonspecific DNA.[63] The interaction of Myc with nonspecific DNA promotes sliding and hopping of the protein whereby the protein can scan for specific DNA sites as observed in prokaryotes.

In contrast to the results discussed above, another study with bacterially produced v-Myc protein localized the DNA binding domain to the carboxy-terminal region corresponding to the basic-helix-loop-helix domain.[64,65] In this study, the DNA binding activity was exclusively determined by calf thymus DNA cellulose chromatography which may favor specific DNA binding under the conditions used.

7.7 HELIX-LOOP-HELIX-LEUCINE ZIPPER OLIGOMERIZATION DOMAIN

The c-Myc carboxy-terminal region is comprised of two contiguous oligomerization motifs, the helix-loop-helix and the leucine zipper. The existence of potential oligomerization domains at the Myc carboxyl-terminus was first recognized with the discovery of the leucine zipper motif (Fig. 7.1) by Landschulz, Johnson and McKnight.[66] C/EBPα, the prototypic leucine zipper protein, was found to have a domain that is highly homologous to the c-Myc carboxy-terminal domain. This domain of c-Myc was previously recognized by Vogt, Bos and Doolittle to be homologous to domains of Fos and the yeast transcriptional activator GCN4 that were predicted to be predominantly alpha-helical.[67] With this realization, Landschulz et al defined a heptad repeat of leucines in C/EBPα, Myc, Fos, and GCN4 that were predicted to align on one side of an alpha helix allowing hydrophobic interactions between two parallel helices, although the original model suggested an antiparallel arrangement of the helices. The leucine zipper is preceded on the N-terminal end with a basic region and thus is called the bZip motif. The bZip dimer is a pair of continuous alpha helices that form a parallel coiled coil over about 30 residues and gradually diverge toward their amino termini to pass through the major groove of the DNA-binding site. The coiled-coil dimerization interface is oriented essentially perpendicular to the DNA axis, giving the protein-DNA complex the appearance of the letter T. This arrangement allows alignment of the contiguous amino-terminal basic regions so they may interact with specific nucleotide sequences within the major groove of DNA. The crystal structures of the GCN4 homodimer and Jun/Fos heterodimer, which are now solved, resemble the expected hypothetical structures.[68-70]

A separate c-Myc region amino-terminal to the leucine zipper was also recognized to be a potential dimerization interface. To determine the significance of the Myc oligomerization domain, Dang et al examined the physicochemical properties of purified recombinant c-Myc proteins produced in bacteria.[71] Gel filtration, velocity sedimentation, and chemical crosslinking experiments indicated that purified Myc protein behaves as a homotetramer that is dependent on the carboxy-terminal 70 amino acids. Deletions that remove either the leucine zipper or helix-loop-helix domains resulted in disruption of tetramerization. Moreover, fusion of the Myc carboxy-terminal 70 amino acids to Staphylococcal protein A, a monomeric protein, caused dimerization of protein A. This observation suggests that these Myc amino acids are sufficient to induce dimerization.

Murre et al later proposed that the helix-loop-helix (HLH) motif in the prototypic HLH protein E2A was also a dimerization domain.[72] The HLH motif consists of a short alpha-helix (helix 1 or H1) linked in parallel fashion to a second short helix (H2) via a loop domain of variable length. This double barrel arrangement allows for hydrophobic interactions between HLH monomers to form a four helix bundle arrangement.[73]

To determine the biological significance of the oligomerization domain of c-Myc, various deletion mutants of c-Myc were assessed for their potential to behave as dominant negative mutants.[71] Dominant negative mutant proteins, as they were originally conceived by Herskowitz, are mutants that lack activity in themselves but are capable of interfering with the wild-type protein through the formation of inactive oligomers with the wild-type protein or inactive hetero-oligomers with partner proteins.[74] Thus, mutations that inactivate the oligomerization domain of c-Myc would not interfere with the wild-type protein since this type of mutant cannot form inactive oligomers with the wild-type protein or potential partner proteins. On the other hand, mutations that inactivate other regions of c-Myc required for transforming, such as the amino-terminal transactivation domain, are expected to result in proteins that can form inactive oligomers with the wild-type protein or with heterooligomeric partner protein. Indeed, studies of such c-Myc mutants indicate that the c-Myc carboxy-terminal region serves as an oligomerization domain. Deletion mutants of c-Myc, N-Myc and L-Myc all behave in a dominant negative manner suggesting that they all dimerize through a common partner protein and share similar transformation pathways.[75]

Studies with bacterially produced Myc protein suggested that c-Myc homo-oligomerizes in vitro.[71] Subsequent studies, however, indicated that homo-oligomerization of c-Myc occurs only at high concentrations of purified proteins in vitro.[76-78] c-Myc does not appear to homo-oligomerize in vivo as initially determined by an adaptation of the two-hybrid system originally conceived in yeast to detect protein-protein interaction in mammalian cells.[71] In such an assay, hybrid proteins constructed from portions of Myc fused to the yeast transcriptional activator GAL4 DNA-binding domain or to the *Herpes simplex* VP16 protein transcriptional activation domain were coexpressed in cells along with a reporter plasmid that responds to a functional GAL4 activator. Neither the separated DNA-binding domain nor the activation domain hybrids alone are capable of activating transcription through GAL4 specific DNA binding sites. Chimeric proteins consisting of these domains fused to heterologous protein segments that are capable of dimerization can bridge the GAL4 DNA-binding domain to the activation domain to reconstitute the transcription factor and activate transcription. In this system, in vivo homo-oligomerization of c-Myc was undetectable, whereas oligomerization between the Fos and Jun leucine zippers was readily dectected. When the c-Myc leucine zipper was replaced by that of the yeast transcriptional activator GCN4, homo-oligomerization between the Myc carboxy-terminal ends were then detectable in vivo. These results suggested that c-Myc functions through hetero-oligomerization with another protein. In fact, Blackwood and Eisenman used the Myc oligomerization domain to functionally clone a partner protein encoded in an expression phage library.[79] This protein, termed Max, is also a helix-loop-helix protein, which will be further discussed in chapter 8.

c-Myc belongs to the bHLHZip (basic-helix-loop-helix-leucine zipper) family of transcriptional regulatory proteins. The bHLHZip motif in the Myc partner protein Max has now been solved through X-ray crystallography.[80] Dimerization between two HLHZip monomers occurs through the HLH domain with the second helix (H2) being continuous with the zipper. This long H2-Zip helix results in extended coiled-coil interaction between two monomers. In addition, the two H1 helices interact in a fashion to form a four helix bundle with the second helices. The basic region extends as a helix continuous with H1. The crystal structures of the HLH domains of E47 (which is related to E12) and MyoD have also solved.[81,82] The arrangements of the helices are comparable, but not identical to those of Max. Numerous mutagenesis studies provided data that are compatible with the crystallographically derived model, however, these hypothetical models diverge from the empirically derived structure in a number of important details, such as the relative positioning of the loop and helices.

7.8 SPECIFIC DNA-BINDING DOMAIN

The amino acid homology of the Myc basic region with other related transcriptional factors was first noticed by Prendergast and Ziff and was proposed to confer specific DNA binding activity.[83] The homology of the Myc basic region with that of the ubiquitous transcription factor USF, which binds a core hexanucleotide sequence CACGTG, led to the identification of a Myc DNA binding site.[84] The c-Myc basic region was tested for its ability to bind DNA by attaching it to the HLH dimerization interface of the E12 enhancer binding factor. Dimers of the chimeric protein specifically bound an E-box element (GGCCACGTGACC) recognized by other HLH proteins in a manner dependent on the integrity of the c-Myc basic motif. The chimeric protein could suppress the cooperation of c-*myc* with H-*ras* in a rat embryo fibroblast transformation assay, suggesting that the DNA recognition specificity of the chimeric protein is related to that of c-Myc in vivo. Through the use of a fusion glutathione-S-transferase-Myc protein and selection of random sequences flanking the consensus CANNTG, Blackwell et al identified the consensus binding site for Myc as CACCACGTGGTG.[78] The selection of this site out of a pool of randomized sequences suggested a specific interaction between Myc and the middle two nucleotides of the core hexamer.

The basic region of c-Myc resembles those of many other basic-helix-loop-helix (bHLH) proteins. Inspection of the linear array of amino acids within the basic region reveals two clusters of basic residues termed basic region 1 (BR1) and basic region 2 (BR2) (Fig. 7.2). bHLH proteins bind the core consensus DNA hexanucleotide sequence, CANNTG, which is called an E-box.[72] In general, the bHLH proteins fall into two classes that differ by their consensus binding sites. Class A proteins bind the core sequence CAGCTG, whereas the class B proteins bind CACGTG.[85] Recognition of these two classes of bHLH proteins led to the hypothesis that the basic regions of proteins within these groups must share common features. On the other hand, there must be amino acid residues that distinguish the class A proteins from those of class B. Figure 7.3 shows the common residues within class A versus class B proteins.

Note that three residues (His, hydrophobic and Arg) are unique to class B proteins which bind CACGTG. Dang and coworkers hypothesized that one or several of these residues in combination contribute to the discrimination between two related DNA hexanucleotide core binding sites.[85] To test this hypothesis, a yeast centromere binding protein and transcription factor, CBF1 or CP1, were utilized. The CBF1 protein functions as a transactivator that regulates the expression of methionine biosynthesis genes in *Saccharomyces cerevisiae*. Thus, loss of CBF1 by homologous recombination in *S. cerevisiae* results in the dependence on methionine (auxotrophy) for growth. The experimental strategy to study CBF1 interaction with specific DNA took advantage of the ability of CBF1 to rescue mutant yeast lacking CBF1 from methionine auxotrophy. Functional CBF1 allows mutant yeast to grow on medium lacking methionine.

The basic region of CBF1 was deleted or replaced with basic regions from class A or class B bHLH proteins. These chimeric activators were tested for their ability to rescue mutant yeast from methionine auxotrophy. While a class A basic region from the transcription factor AP4 could not function, several class B basic regions allowed the mutant yeast strain to grow in the absence of methionine. These results implied that class B basic regions may

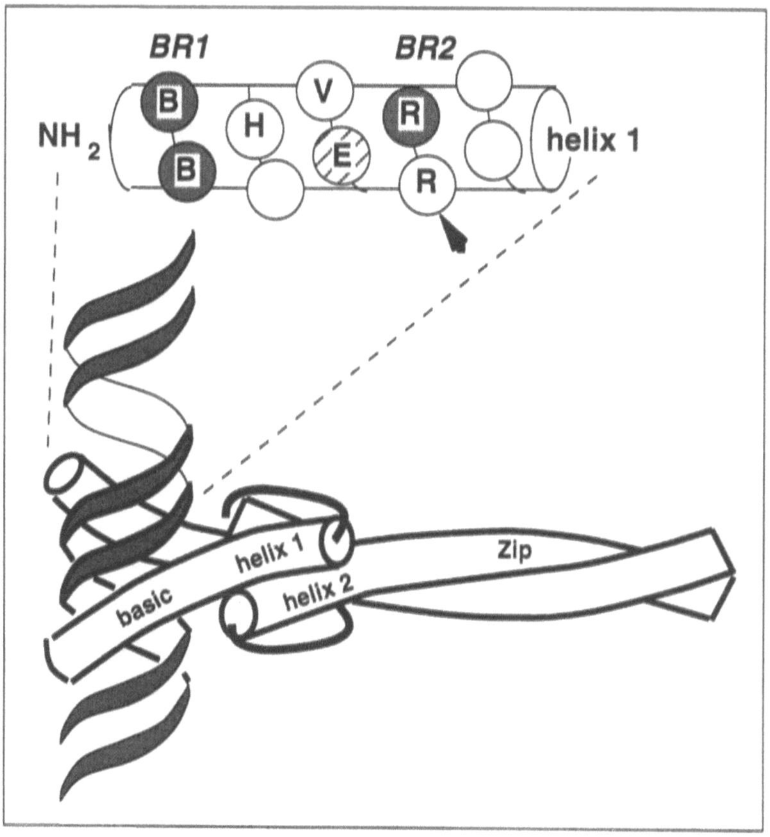

Fig. 7.2. c-Myc protein basic domain. BR1 and BR2 designate clusters of basic residues. The dimerization of bHLH proteins allows them to bind DNA by aligning the basic regions in the major groove of DNA. The arrow indicates the arginine at position 367 that is important for the discrimination of the similar DNA binding sites.

be functionally equivalent. To identify the amino acid residues responsible for discriminating between the core binding sites CA<u>CG</u>TG and CA<u>GC</u>TG, the inactive CBF1-AP4 chimeric protein containing the AP4 basic region was mutated at the three amino acids that distinguish class A from class B proteins. By substituting only the Arg residue (in the sequence ERR<u>R</u>), the CBF1-AP4 chimeric protein was endowed with the ability to rescue methionine auxotrophy. Furthermore, randomization of codons at this position also revealed that alleles capable of reconstituting CBF1 function encode Arg and in several cases Lys at that position. Independently, Prochownik and coworkers were able to alter the specificity of MyoD DNA binding by mutating a single amino acid within the basic region which corresponds to the same position as the conserved Arg in the Myc family of protein.[86] Thus, a single amino acid residue, Arg, within the basic region of bHLH proteins appears to be critical for discriminating between two similar DNA sites.

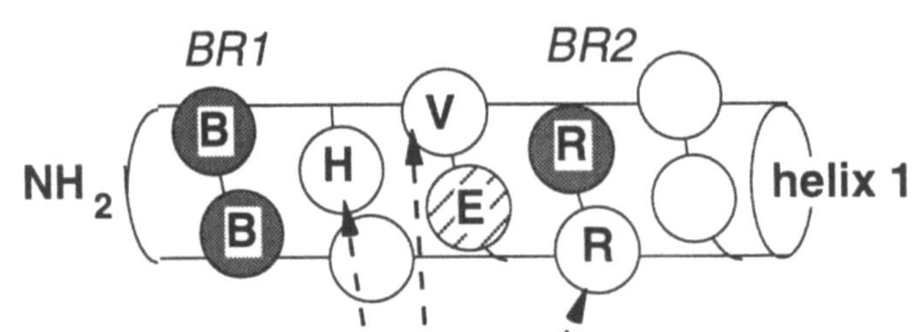

BASIC REGIONS — BINDING SITE

	bHLH	--	BB	--N--	ER-R	X	--w---w--	L	--w	BINDING SITE
Class A	AP4	RI	RR	EIANSN	ERRR	M	QSINAGFQS	L	KTL	
	MyoD	AD	RR	KAATMR	ERRR	L	SKVNEAFET	L	KRC	CAG CTG
	E12	KE	RR	VANNAR	ERLR	V	RDINEAFKE	L	GRM	
	E47	RE	RR	MANNAR	ERVR	V	RDINEAFRE	L	GRM	
	Tal1	VV	RR	IFTNSR	ERWR	Q	QNVNGAFAE	L	RKL	
				* *		*				
Class B	c-Myc	NV	KR	RT**H**NV**L**	ERQR	**R**	NELKRSFFA	L	RDQ	
	USF	EK	RR	AQ**H**NE**V**	ERRR	**R**	DKINNWIVQ	L	SKI	CAC GTG
	CBF1	KQ	RK	DS**H**KE**V**	ERRR	**R**	ENINTAINV	L	SDL	
	Max	AD	KR	AH**H**NA**L**	ERRR	**R**	DHIKDSFHS	L	RDS	
	TFE3	RQ	KK	DN**H**NL**I**	ERRR	**R**	FNINDRIKE	L	GTL	
	PHO4	DD	<u>KR</u>	ES**H**KH**A**	<u>EQAR</u>	**R**	NRLAVALHE	L	ASL	
			BR1		BR2					

Fig. 7.3. Two classes of bHLH proteins. The class A and B proteins are distinguished by the DNA binding site each recognizes. The basic region of the bHLH proteins depicted as a helix is responsible for specific DNA binding. Class B proteins possess three unique residues, histidine, hydrophobic, and arginine (shown in bold), which permit them to recognize an alternate DNA binding site than that recognized by the Class A proteins.

Photocrosslinking studies using site-specific 5-bromouracil indicate that the conserved His residue in class B contacts the thymine in CANN<u>T</u>G.[87] Recent X-ray crystallographic structures of dimeric Max indicate that the homologous His residue in Max also appears very close to the thymine residue, although the Max His residue forms a hydrogen bond with the outer guanine of the G:C pairs.[80] The conserved Arg residue of Max, which is found in class B proteins, contacts the guanine residue of the middle two nucleotides: CAC<u>G</u>TG. In the case of class A proteins, the crystal structures of MyoD and E2A did not show consistent interactions between specific amino acids and the middle two nucleotides.[81,82]

Current crystal structures of bHLH proteins do not resolve the interactions between protein and sequences flanking the core hexanucleotide, CANNTG.[80,81,82] Studies of flanking sequences, however, indicate preferences of specific bHLH proteins for certain nucleotides flanking the core hexanucleotide (see below, DNA-binding properties of Myc). Thus, other amino acid residues are determinants for specific interactions with flanking nucleotides. The available crystal structures do not reveal details outside of the hexanucleotide core CANNTG leaving it possible that amino acids in the loops may influence flanking site preferences. Differences in flanking site preferences may be highly physiologically relevant for identification of target genes of bHLH proteins. Additional studies are necessary to delineate interactions between bHLH proteins and nucleotides flanking the hexanucleotide core sequence.

7.9 TRANSCRIPTIONAL ACTIVATION DOMAIN

Localization of c-Myc to the nucleus suggests that it may function in some known nuclear process, such as DNA replication, transcription or post-transcriptional regulation. In fact, participation in all three possible processes has been ascribed to c-Myc, although its role in transcriptional regulation has held up best to intense scrutiny. An early study in yeast indicated that full length c-Myc fused to the Lex A DNA-binding domain was capable of weakly transactivating a Lex A dependent reporter gene as compared to Fos tethered to Lex A.[88] A later study sought to determine whether c-Myc sequences are able to activate transcription in vertebrate cells using various regions of c-*myc* fused to sequences encoding the yeast GAL4 DNA binding domain.[89]

It became readily apparent that the amino-terminal 143 amino acids of c-Myc are able to activate transcription of a GAL4 dependent reporter gene construct (Fig. 7.1).[89] This region was further dissected into three arbitrary subregions via available restriction sites: region A (aa. 1-41); region B (aa. 41-100); and region C (aa. 103-143). The subregions independently activate transcription when tethered to GAL4. It was noted that the amino-terminal 143 amino acids are also those which were previously shown to be required for neoplastic transformation. Curiously, deletion of amino acids 106-143 abrogated transforming activity in full length c-Myc, but removal of this region did not reduce the transactivation activity of the chimeric GAL4 protein.

Whether v-Myc protein utilizes a similar activation domain was determined.[90,91] The MC29 *gag-myc* sequences were fused to GAL4 to determine whether chimeric GAL4 proteins could activate transcription of a GAL4 dependent reporter. The amino-terminal region of v-Myc tethered to GAL4 effectively activates transcription, whereas the full length protein (-32-416) failed to activate.[90] Similarly, inclusion of the carboxy-terminal oligomerization region inhibited the c-Myc transactivation domain. Subregions in v-Myc capable of transactivation were also identified: aa. -33-42 and 90-219. Deletions of these regions in v-Myc led to a reduction in transforming activity.[6,90] These observations suggest a correlation between transactivation and transformation by Myc.

Further proof of a correlation between transactivation and transformation activities of c-Myc were gained from studies of related proteins, L-Myc and B-Myc. L-Myc, like c-Myc, is able to transform rat embryo cells in cooperation with an activated Ras, although L-Myc is much less efficient than c-Myc. The amino-terminal region of L-Myc is also capable of activating transcription when tethered to GAL4, but only at about 10% of that of c-Myc.[92] Chimeras containing swapped amino-terminal domains of c-Myc and L-Myc revealed that the c-Myc transactivation domain, which is more potent in transcription assays, confers more effective transforming activity. These studies suggest that the potency of c-Myc as compared to L-Myc in transformation assays correlates with its amino-terminal domain.

B-Myc is a unique protein that contains sequences corresponding to the c-Myc transactivation domain and lacks sequences suggestive of a DNA binding domain.[93-95] To study the potential function of B-Myc, it was hypothesized that B-Myc is able to squelch c-Myc transactivation function.[16] B-Myc, when tethered to GAL4, is more potent than c-Myc in activating transcription. Moreover, titration experiments suggest that B-Myc is capable of inhibiting c-Myc transactivation and transforming activity. These studies again support a link between transactivation and transformation.

Studies with full length c-Myc have further indicated that the amino-terminal domain supplies transactivation activity.[96-100] The carboxy-terminal region has been well established to mediate an interaction with Max. Deletions of the amino-terminal c-Myc region resulted in the loss of transactivation activity of Myc-Max complexes. Thus E-box dependent transcription relies on both the DNA binding activity of the Myc-Max complex through the bHLHZip domain and the amino-terminal transactivation domain.

It has become apparent that c-Myc transforming activity is not purely related to its ability to activate transcription. Deletion of the c-Myc transactivation domain region including amino acids 122-144 (Myc box II) is associated with the loss of transcriptional suppression through the initiator sequence (see chapter 10) and abrogation of transforming activity.[41] This deletion, however, does not abolish the transactivation function of c-Myc. Moreover, chimeras of c-Myc containing the potent *Herpes simplex* VP16 transactivation domain do not transform cells.[17] Recently the Myc box II region has been shown to be required for an interaction with a nuclear factor that also interacts functionally with E1A.[17] An alteration of Myc box II that results in the loss of binding to this factor is associated with the loss of transforming activity. Although the identity of this putative Myc interacting factor is still elusive, its eventual isolation will likely shed significant insight into the transforming and transcriptional regulatory functions of c-Myc. Thus, the transactivation domain of c-Myc is a complex region composed of subdomains that are likely to serve distinct functions.

7.10 MYC, TBP AND YY1

While it has been reproducibly demonstrated that the c-Myc amino-terminal domain is able to activate transcription of reporter plasmids, the mechanisms by which c-Myc interacts with the transcriptional machinery remain virtually unknown. Although a specific association of c-Myc and TATA binding protein-associated factors (TAFs) has not been detected, several studies suggest that c-Myc is able to interact directly with TATA binding protein (TBP).[101,102] The c-Myc transactivation domain interacts with TBP in vitro. Crosslinking studies performed in situ indicate that c-Myc directly interacts with TBP in vivo. In addition, results from recent two-hybrid experiments also suggest a specific interaction between the c-Myc transactivation domain and TBP in vivo.[103] Despite these observations, there is reasonable skepticism regarding the physiological relevance of c-Myc interaction with TBP as TBP appears to be generally 'sticky'. On the other hand, this interaction should not so readily be dismissed as artifactual, since direct interactions between transcription factors and TBP have been demonstrated. Thus, genetic as well as in vitro reconstitution experiments are necessary to enhance our appreciation of the mechanisms underlying c-Myc transactivation.

Clues to these mechanisms of c-Myc-mediated transformation may also come from insight into the interactions of novel proteins with c-Myc. For example, YY1, a transcription factor known to associate with certain initiator elements and to activate the c-*myc* promoter, was found to interact with c-Myc in a two-hybrid screen.[104,105] In transfection experiments, c-Myc inhibits the ability of YY1 to initiate transcription of target reporter plasmids. How YY1 participates in c-Myc mediated phenotypes like cell growth or apoptosis remains unknown.

7.11 INTERACTION OF p107 AND MYC

The retinoblastoma-related protein, p107, binds to the c-Myc amino-terminal transactivation domain.[103,106,107] By using a radiolabeled fusion Myc amino-terminal protein to screen an expression library for cDNAs encoding Myc binding proteins, Gu et al cloned a partial cDNA that corresponds to the pocket domain of the p107 protein.[106] Earlier unpublished studies by Bernards and coworkers (1993 Oncogenes meeting, Frederick, MD) indicated that p107 is able to inhibit GAL4-Myc transactivation whereas p105 (pRb) could not, conflicting with a previous report that pRb directly interacts with the Myc amino-terminal domain.[108]

p107 inhibited transactivation by GAL4-Myc, but could not effectively inhibit GAL4 fused to mutated Myc derived from Burkitt's lymphomas.[103,106] Virtually all of the somatic mutations of Myc are confined to the first 200 amino-terminal residues (Fig. 5.2). The inhibition of Myc by p107 appears to be selective since p107 did not inhibit GAL4 fused to other transactivation domains such as those of VP16, USF (Dolde and Dang, unpublished observation) or NF1/CTF1. The mechanisms underlying the inability of p107 to inhibit c-Myc mutants were investigated. The most elegant hypothesis considered was that mutations in the Myc activation domain would simply interfere with the binding of p107 and Myc. Results from in vitro binding studies, mammalian two-hybrid assays and immunoprecipitation experiments suggest that p107 associates with wild-type and mutant c-Myc equivalently.[103] How then does p107 differentially inhibit wild-type and mutant Myc transactivation?

Clues to the differential effect of p107 were provided by the sites of mutation in c-Myc. Inspection of the distribution of point mutations in the c-Myc transactivation domain found in Burkitt's and AIDS-related lymphoma as well as those alterations found in v-Myc indicate that the majority of the mutations cluster around two known phosphorylation sites, Thr-58 and Ser-62 that is within a consensus CDK phosphorylation site (Fig. 7.4).[109-111] In addition, p107 was shown to directly associate with either cyclin A or E and the associated CDKs.[112-114] It was envisioned that p107 associates with the c-Myc activation domain and carries with it the cyclin A/E/CDK complexes which are able to phosphorylate and inactivate c-Myc transactivation function.

Biologically, this model suggests a negative feedback loop whereby c-Myc function is downregulated in S/G_2 phases by phosphorylation. To first test this hypothesis, the ability of p107/cyclin A/CDK complexes to phosphorylate c-Myc in vitro was determined.[103] c-Myc was phosphorylated at Ser-62 by p107/CDK. In a crude reconstitution phosphorylation assay in reticulocyte lysate, both Ser-62 and Thr-58 were phosphorylated in a cyclin A dependent manner suggesting that cyclin A/CDK is able to initiate sequential phosphorylation reactions on c-Myc (Fig. 7.4). The kinase responsible for Thr-58 phosphorylation has not been definitively identified, although GSK3β could phosphorylate Thr-58 in vitro. As hypothesized, recombinant c-Myc proteins containing lymphoma-derived mutations flanking either Thr-58 or Ser-62 were unable to be phosphorylated at Thr-58 in a cyclin A dependent manner (Fig. 7.5).

Although these experiments provide a reasonable explanation for the differential effects of p107 on c-Myc and c-Myc mutants, additional evidence regarding the physiological interaction between p107 and Myc is required to substantiate the hypothesis outlined above. For example, will a cell cycle specific association between Myc and p107 be detected? Other questions abound from recent observations that mutation of Thr-58 augments c-Myc transforming activity as well as past recognition that Thr-58 is invariably mutated in v-Myc. What is the kinase that phosphorylates Thr-58? There are also conflicting results regarding the influence of these phosphorylation sites on the ability of c-Myc to transactivate; additional studies are required to resolve these issues.

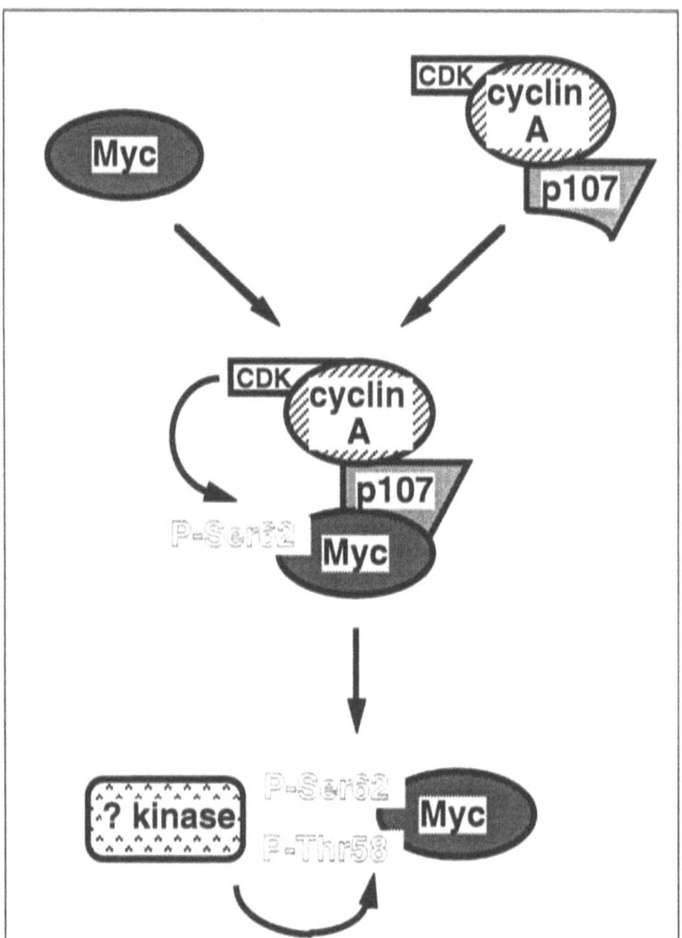

Fig. 7.4. Model of sequential phosphorylation of c-Myc residues Ser-62 and Thr-58. The p107-cyclin A-CDK complex putatively phosphorylates c-Myc at Ser-62 in vivo allowing the subsequent phosphorylation of Thr-58 by an as yet unidentified kinase.

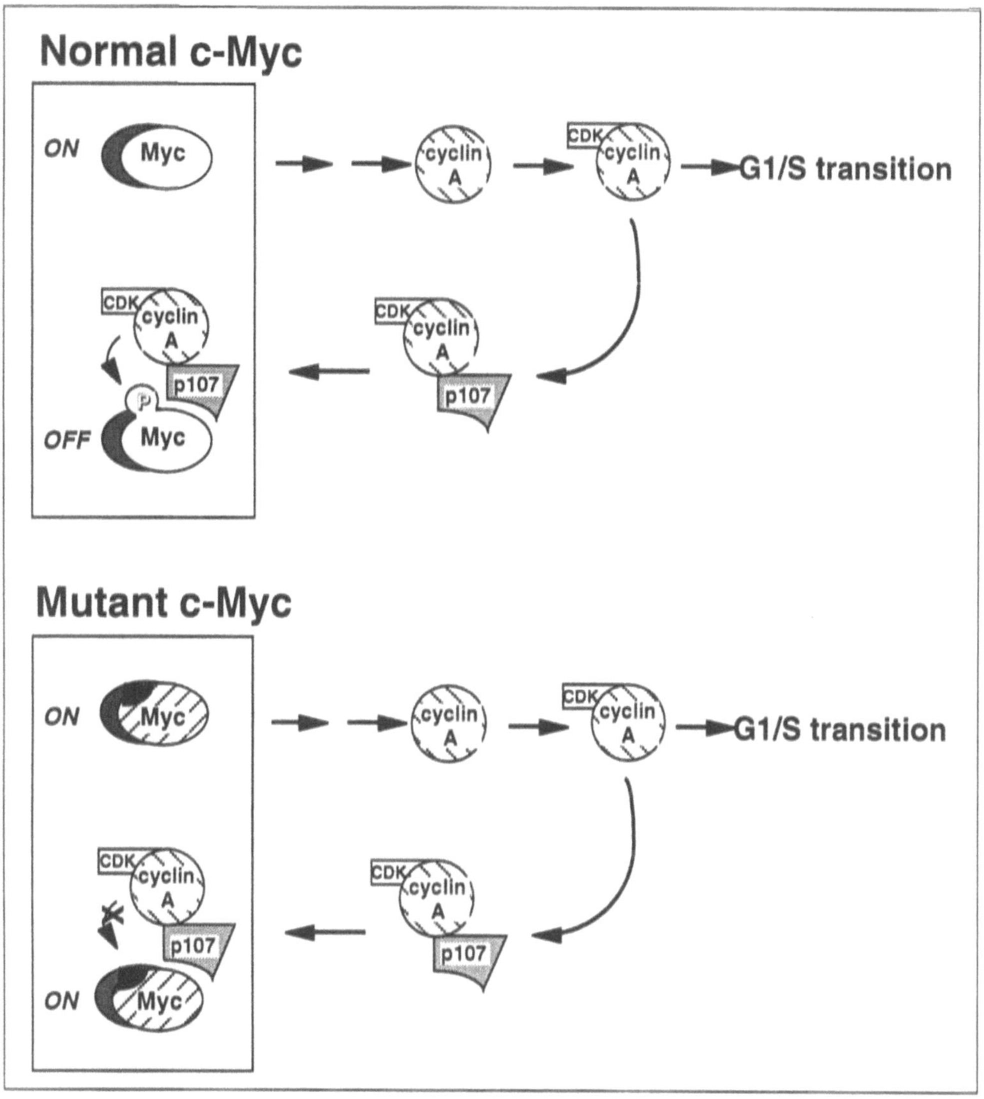

Fig. 7.5. Model of p107 suppression of Myc transactivation. A. Myc is thought to effect G1/S phase transition by upregulating the expression of cyclin A. Top: Cyclin A forms a complex with its cyclin-dependent kinase. The complex interacts with p107, leading ultimately to phosphorylation of Myc in its transactivation domain, and rendering it incapable of transactivation. Below: Lymphoma-derived mutant alleles of myc *contain mutations at or near the cyclin A-CDK phosphorylation site within the transactivation domain, such that Myc transactivation cannot be inhibited by p107.*

REFERENCES

1. Land H, Parada LF, Weinberg RA. Tumorigenic conversion of primary embryo fibroblasts requires at least two cooperating oncogenes. Nature 1983; 304:596-602.
2. Land H, Parada LF, Weinberg RA. Cellular oncogenes and multistep carcinogenesis. [Review]. Science 1983; 222:771-8.
3. Small MB, Hay N, Schwab M, Bishop JM. Neoplastic transformation by the human gene N-myc. Mol Cell Biol 1987; 7:1638-1645.
4. Stone J, de Lange T, Ramsay G, et al. Definition of regions in human c-myc that are involved in transformation and nuclear localization. Mol Cell Biol 1987; 7:1697-709.
5. Bister K, Trachmann C, Jansen HW, Schroeer B, Patschinsky T. Structure of mutant and wild-type MC29 v-myc alleles and biochemical properties of their protein products. Oncogene 1987; 1:97-109.
6. Farina SF, Huff JL, Parsons JT. Mutations within the 5' half of the avian retrovirus MC29 v-myc gene alter or abolish transformation of chicken embryo fibroblasts and macrophages. J Virol 1992; 66:2698-708.
7. Sarid J, Halazonetis TD, Murphy W, Leder P. Evolutionarily conserved regions of the human c-myc protein can be uncoupled from transforming activity. Proc Natl Acad Sci USA 1987; 84:170-3.
8. Heaney ML, Pierce J, Parsons JT. Site-directed mutagenesis of the gag-myc gene of avian myelocytomatosis virus 29: biological activity and intracellular localization of structurally altered proteins. J Virol 1986; 60:167-76.
9. Biegalke BJ, Heaney ML, Bouton A, Parsons JT, Linial M. MC29 deletion mutants which fail to transform chicken macrophages are competent for transformation of quail macrophages. J Virol 1987; 61:2138-42.
10. Freytag SO, Dang CV, Lee WM. Definition of the activities and properties of c-myc required to inhibit cell differentiation. Cell Growth Differ 1990; 1:339-43.
11. Bar-Ner M, Messing LT, Segal S. Inhibition of murine erythroleukemia cell differentiation by normal and partially deleted c-myc genes. [Review]. Immunobiol 1992; 185:150-8.
12. Bar-Ner M, Messing LT, Cultraro CM, Birrer MJ, Segal S. Regions within the c-Myc protein that are necessary for transformation are also required for inhibition of differentiation of murine erythroleukemia cells. Cell Growth Differ 1992; 3:183-90.
13. Ohmori Y, Tanabe J, Takada S, Lee WM, Obinata M. Functional domains of c-Myc involved in the commitment and differentiation of murine erythroleukemia cells. Oncogene 1993; 8:379-86.
14. Prochownik EV, Kukowska J. Deregulated expression of c-myc by murine erythroleukaemia cells prevents differentiation. Nature 1986; 322:848-50.

15. Takada S, Yamamoto T, Ohmori Y, Matsui Y, Obinata M. c-Myc interferes with the commitment to differentiation of murine erythroleukemia cells at a reversible point. Jpn J Cancer Res 1992; 83:61-5.
16. Resar LM, Dolde C, Barrett JF, Dang CV. B-myc inhibits neoplastic transformation and transcriptional activation by c-myc. Mol Cell Biol 1993; 13:1130-6.
17. Brough DE, Hofmann TJ, Ellwood KB, Townley RA, Cole MD. An essential domain of the c-myc protein interacts with a nuclear factor that is also required for E1A-mediated transformation. Mol Cell Biol 1995; 15:1536-44.
18. Kaczmarek L, Hyland JK, Watt R, Rosenberg M, Baserga R. Microinjected c-myc as a competence factor. Science 1985; 228:1313-5.
19. Goruppi S, Gustincich S, Brancolini C, Lee WM, Schneider C. Dissection of c-myc domains involved in S phase induction of NIH3T3 fibroblasts. Oncogene 1994; 9:1537-44.
20. Studzinski GP, Shankavaram UT, Moore DC, Reddy PV. Association of c-myc protein with enzymes of DNA replication in high molecular weight fractions from mammalian cells. J Cell Physiol 1991; 147:412-9.
21. Studzinski GP, Brelvi ZS, Feldman SC, Watt RA. Participation of c-myc protein in DNA synthesis of human cells. Science 1986; 234:467-70.
22. Gutierrez C, Guo ZS, Farrell-Towt J, Ju G, DePamphilis ML. c-myc protein and DNA replication: separation of c-myc antibodies from an inhibitor of DNA synthesis. Mol Cell Biol 1987; 7:4594-8.
23. Gutierrez C, Guo ZS, Burhans W, DePamphilis ML, Farrell-Towt J, Ju G. Is c-myc protein directly involved in DNA replication? Science 1988; 240:1202-3.
24. Iguchi-Ariga SM, Itani T, Kiji Y, Ariga H. Possible function of the c-myc product: promotion of cellular DNA replication. EMBO J 1987; 6:2365-71.
25. Iguchi-Ariga SM, Itani T, Yamaguchi M, Ariga H. c-myc protein can be substituted for SV40 T antigen in SV40 DNA replication. Nucleic Acids Res 1987; 15:4889-99.
26. Iguchi-Ariga SM, Okazaki T, Itani T, Ogata M, Sato Y, Ariga H. An initiation site of DNA replication with transcriptional enhancer activity present upstream of the c-myc gene. EMBO J 1988; 7:3135-42.
27. Ariga H, Imamura Y, Iguchi-Ariga SM. DNA replication origin and transcriptional enhancer in c-myc gene share the c-myc protein binding sequences. EMBO J 1989; 8:4273-9.
28. Berberich S, Trivedi A, Daniel DC, Johnson EM, Leffak M. In vitro replication of plasmids containing human c-myc DNA. J Mol Biol 1995; 245:92-109.

29. Classon M, Henriksson M, Sumegi J, Klein G, Hammaskjold ML. Elevated c-myc expression facilitates the replication of SV40 DNA in human lymphoma cells. Nature 1987; 330:272-4.
30. Classon M, Wennborg A, Henriksson M, Klein G. Analysis of c-Myc domains involved in stimulating SV40 replication. Gene 1993; 133:153-61.
31. Classon M, Henriksson M, Klein G, Sumegi J. The effect of myc proteins on SV40 replication in human lymphoid cells. Oncogene 1990; 5:1371-6.
32. Tribioli C, Biamonti G, Giacca M, Colonna M, Riva S, Falaschi A. Characterization of human DNA sequences synthesized at the onset of S-phase. Nucleic Acids Res 1987; 15:10211-32.
33. Falaschi A, Biamonti G, Cobianchi F, et al. Presence of transcription signals in two putative DNA replication origins of human cells. Biochim Biophys Acta 1988; 951:430-42.
34. Giacca M, Gutierrez MI, Demarchi F, et al. A protein target site in an early replicated human DNA sequence: a highly conserved binding motif. Biochem Biophys Res Commun 1989; 165:956-65.
35. DePamphilis ML. Eukaryotic DNA replication: anatomy of an origin. [Review]. Annu Rev Biochem 1993; 62:29-63.
36. DePamphilis ML. Origins of DNA replication in metazoan chromosomes. [Review]. J Biol Chem 1993; 268:1-4.
37. DePamphilis ML. Origins of DNA replication that function in eukaryotic cells. [Review]. Curr Opin Cell Biol 1993; 5:434-41.
38. Serizawa H, Makela TP, Conaway JW, Conaway RC, Weinberg RA, Young RA. Association of Cdk-activating kinase subunits with transcription factor TFIIH. Nature 1995; 374:280-2.
39. Penn LJ, Brooks MW, Laufer EM, Land H. Negative autoregulation of c-myc transcription. Embo J 1990; 9:1113-21.
40. Grignani F, Lombardi L, Inghirami G, Sternas L, Cechova K, Dalla-Favera R. Negative autoregulation of c-myc gene expression is inactivated in transformed cells. EMBO J 1990; 9:3913-22.
41. Li LH, Nerlov C, Prendergast G, MacGregor D, Ziff EB. c-Myc represses transcription in vivo by a novel mechanism dependent on the initiator element and Myc box II. EMBO J 1994; 13:4070-9.
42. Buckle RS, Mechali M. Analysis of c-Myc and Max binding to the c-myc promoter: evidence that autosuppression occurs via an indirect mechanism. Oncogene 1995; 10:1249-55.
43. Penn LJ, Brooks MW, Laufer EM, et al. Domains of human c-myc protein required for autosuppression and cooperation with ras oncogenes are overlapping. Mol Cell Biol 1990; 10:4961-6.
44. Persson H, Leder P. Nuclear localization and DNA binding properties of a protein expressed by human c-myc oncogene. Science 1984; 225:718-21.
45. Hann SR, Eisenman RN. Proteins encoded by the human c-myc oncogene: differential expression in neoplastic cells. Mol Cell Biol 1984; 4:2486-97.

46. Ramsay G, Evan GI, Bishop JM. The protein encoded by the human proto-oncogene c-myc. Proc Natl Acad Sci USA 1984; 81:7742-6.
47. Eisenman RN, Tachibana CY, Abrams HD, Hann SR. V-myc- and c-myc-encoded proteins are associated with the nuclear matrix. Mol Cell Biol 1985; 5:114-26.
48. Gusse M, Ghysdael J, Evan G, Soussi T, Mechali M. Translocation of a store of maternal cytoplasmic c-myc protein into nuclei during early development. Mol Cell Biol 1989; 9: 5395-403.
49. Vriz S, Lemaitre JM, Leibovici M, Thierry N, Mechali M. Comparative analysis of the intracellular localization of c-Myc, c-Fos, and replicative proteins during cell cycle progression. Mol Cell Biol 1992; 12:3548-55.
50. Boulikas T. Nuclear localization signals (NLS). [Review]. Crit Rev Eukaryot Gene Expr 1993; 3:193-227.
51. Dang CV, Lee WM. Identification of the human c-myc protein nuclear translocation signal. Mol Cell Biol 1988; 8:4048-54.
52. Tikhonenko AT, Hartman AR, Linial ML. Overproduction of v-Myc in the nucleus and its excess over Max are not required for avian fibroblast transformation. Mol Cell Biol 1993; 13:3623-31.
53. Kato GJ, Wechsler DS, Dang CV. DNA binding by the Myc oncoproteins. [Review]. Cancer Treat Res 1992; 63:313-25.
54. Bunte T, Donner P, Pfaff E, et al. Inhibition of DNA binding of purified p55v-myc in vitro by antibodies against bacterially expressed myc protein and a synthetic peptide. EMBO J 1984; 3:1919-24.
55. Watt RA, Shatzman AR, Rosenberg M. Expression and characterization of the human c-myc DNA-binding protein. Mol Cell Biol 1985; 5:448-56.
56. Bunte T, Greiser-Wilke I, Donner P, Moelling K. Association of gag-myc proteins from avian myelocytomatosis virus wild-type and mutants with chromatin. EMBO J 1982; 1:919-27.
57. Bunte T, Greiser-Wilke I, Moelling K. The transforming protein of the MC29-related virus CMII is a nuclear DNA-binding protein whereas MH2 codes for a cytoplasmic RNA-DNA binding polyprotein. EMBO J 1983; 2:1087-92.
58. Beimling P, Benter T, Sander T, Moelling K. Isolation and characterization of the human cellular myc gene product. Biochemistry 1985; 24:6349-55.
59. Donner P, Bunte T, Greiser-Wilke I, Moelling K. Decreased DNA-binding ability of purified transformation-specific proteins from deletion mutants of the acute avian leukemia virus MC29. Proc Natl Acad Sci USA 1983; 80:2861-5.
60. Dang CV, van Dam H, Buckmire M, Lee WMF. DNA-binding domain of human c-myc produced in Escherichia coli. Mol Cell Biol 1989; 9:2477-2486.

61. Suzuki M. SPXX, a frequent sequence motif in gene regulatory proteins. J Mol Biol 1989; 207:61-84.
62. Churchill ME, Suzuki M. 'SPKK' motifs prefer to bind to DNA at A/T-rich sites. EMBO J 1989; 8:4189-95.
63. von Hippel PH, Berg OG. Facilitated target location in biological systems. [Review]. J Biol Chem 1989; 264:675-8.
64. Kerkhoff E, Bister K. Myc protein structure: localization of DNA-binding and protein dimerization domains. Oncogene 1991; 6:93-102.
65. Kerkhoff E, Bister K, Klempnauer KH. Sequence-specific DNA binding by Myc proteins. Proc Natl Acad Sci USA 1991; 88:4323-7.
66. Landschulz WH, Johnson PF, McKnight SL. The leucine zipper: a hypothetical structure common to a new class of DNA binding proteins. Science 1988; 240:1759-64.
67. Vogt PK, Bos TJ, Doolittle RF. Homology between the DNA-binding domain of the GCN4 regulatory protein of yeast and the carboxy-terminal region of a protein coded for by the oncogene jun. Proc Natl Acad Sci USA 1987; 84:3316-9.
68. O'Shea EK, Klemm JD, Kim PS, Alber T. X-ray structure of the GCN4 leucine zipper, a two-stranded, parallel coiled coil. Science 1991; 254:539-44.
69. Ellenberger TE, Brandl CJ, Struhl K, Harrison SC. The GCN4 basic region leucine zipper binds DNA as a dimer of uninterrupted alpha helices: crystal structure of the protein-DNA complex. Cell 1992; 71:1223-37.
70. Glover JN, Harrison SC. Crystal structure of the heterodimeric bZIP transcription factor c-Fos-c-Jun bound to DNA. Nature 1995; 373:257-61.
71. Dang CV, McGuire M, Buckmire M, Lee WMF. Involvement of the'leucine zipper' region in the oligomerization and transforming activity of human c-myc protein. Nature 1989; 337:664-6.
72. Murre C, McCaw PS, Baltimore D. A new DNA binding and dimerization motif in immunoglobulin enhancer binding, daughterless, MyoD, and myc proteins. Cell 1989; 56:777-83.
73. Murre C, McCaw PS, Vaessin H, et al. Interactions between heterologous helix-loop-helix proteins generate complexes that bind specifically to a common DNA sequence. Cell 1989; 58:537-44.
74. Herskowitz I. Functional inactivation of genes by dominant negative mutations. [Review]. Nature 1987; 329:219-22.
75. Mukherjee B, Morgenbesser SD, DePinho RA. Myc family oncoproteins function through a common pathway to transform normal cells in culture: cross-interference by Max and trans-acting dominant mutants. Genes Dev 1992; 6:1480-92.
76. Smith MJ, Charron-Prochownik DC, Prochownik EV. The leucine zipper of c-Myc is required for full inhibition of erythroleukemia differentiation. Mol Cell Biol 1990; 10:5333-9.

77. Littlewood TD, Amati B, Land H, Evan GI. Max and c-Myc/Max DNA-binding activities in cell extracts. Oncogene 1992; 7:1783-92.
78. Blackwell TK, Kretzner L, Blackwood EM, Eisenman RN, Weintraub H. Sequence-specific DNA binding by the c-Myc protein. Science 1990; 250:1149-51.
79. Blackwood EM, Eisenman RN. Max: a helix-loop-helix zipper protein that forms a sequence-specific DNA-binding complex with Myc. Science 1991; 251:1211-7.
80. Ferre-D'Amare AR, Prendergast GC, Ziff EB, Burley SK. Recognition by Max of its cognate DNA through a dimeric b/HLH/Z domain. Nature 1993; 363:38-45.
81. Ellenberger T, Fass D, Arnaud M, Harrison SC. Crystal structure of transcription factor E47 - E-box recognition by a basic region helix-loop-helix dimer. Genes Dev 1994; 8:970-980.
82. Ma P, Rould MA, Weintraub H, Pabot CO. Crystal structure of MyoD bHLH domain-DNA complex - perspectives on DNA recognition and implications for transcriptional activation. Cell 1994; 77:451-9.
83. Prendergast GC, Ziff EB. DNA-binding motif [letter]. Nature 1989; 341:392.
84. Prendergast GC, Ziff EB. Methylation-sensitive sequence-specific DNA binding by the c-Myc basic region. Science 1991; 251:186-9.
85. Dang CV, Dolde C, Gillison ML, Kato GJ. Discrimination between related DNA sites by a single amino acid residue of Myc-related basic-helix-loop-helix proteins. Proc Nat Acad Sci USA 1992; 89:599-602.
86. Van Antwerp ME, Chen DG, Chang C, Prochownik EV. A point mutation in the MyoD basic domain imparts c-Myc-like properties. Proc Nat Acad Sci USA 1992; 89:9010-9014.
87. Dong QP, Blatter EE, Ebright YW, Bister K, Ebright RH. Identification of amino acid base contacts in the myc dna complex by site-specific bromouracil mediated photocrosslinking. EMBO J 1994; 13:200-4.
88. Lech K, Anderson K, Brent R. DNA-bound Fos proteins activate transcription in yeast. Cell 1988; 52:179-84.
89. Kato GJ, Barrett J, Villa-Garcia M, Dang CV. An amino-terminal c-myc domain required for neoplastic transformation activates transcription. Mol Cell Biol 1990; 10:5914-20.
90. Min S, Taparowsky EJ. v-Myc, but not Max, possesses domains that function in both transcription activation and cellular transformation. Oncogene 1992; 7:1531-40.
91. Min SY, Cridermiller SJ, Taparowsky EJ. The transcription activation domains of v-myc and vp16 interact with common factors required for cellular transformation and proliferation. Cell Growth Differ 1994; 5:563-73.

92. Barrett J, Birrer MJ, Kato GJ, Dosaka-Akita H, Dang CV. Activation Domains of L-Myc and c-Myc determine their transforming potencies in rat embryo cells. Mol Cell Biol 1992:3130-7.
93. Asker C, Steinitz M, Andersson K, Sumegi J, Klein G, Ingvarsson S. Nucleotide sequence of the rat B myc gene. Oncogene 1989; 4:1523-7.
94. Ingvarsson S, Sundaresan S, Jin P, et al. Chromosome localization and expression pattern of Lmyc and Bmyc in murine embryonal carcinoma cells. Oncogene 1988; 3:679-85.
95. Ingvarsson S, Asker C, Axelson H, Klein G, Sumegi J. Structure and expression of B-myc, a new member of the myc gene family. Mol Cell Biol 1988; 8:3168-74.
96. Amati B, Dalton S, Brooks MW, Littlewood TD, Evan GI, Land H. Transcriptional activation by the human c-Myc oncoprotein in yeast requires interaction with Max. Nature 1992; 359:423-6.
97. Amati B, Brooks MW, Levy N, Littlewood TD, Evan GI, Land H. Oncogenic activity of the c-Myc protein requires dimerization with Max. Cell 1993; 72:233-45.
98. Amin C, Wagner AJ, Hay N. Sequence-specific transcriptional activation by Myc and repression by Max. Mol Cell Biol 1993; 13:383-90.
99. Kretzner L, Blackwood EM, Eisenman RN. Myc and Max proteins possess distinct transcriptional activities. Nature 1992; 359:426-9.
100. Gu W, Cechova K, Tassi V, Dalla-Favera R. Opposite regulation of gene transcription and cell proliferation by c-Myc and Max. Proc Nat Acad Sci USA 1993; 90:2935-9.
101. Hateboer G, Timmers HT, Rustgi AK, Billaud M, van't Veer LJ, Bernards R. TATA-binding protein and the retinoblastoma gene product bind to overlapping epitopes on c-Myc and adenovirus E1A protein. Proc Natl Acad Sci USA 1993; 90:8489-93.
102. Maheswaran S, Lee H, Sonenshein GE. Intracellular association of the protein product of the c-myc oncogene with the TATA-binding protein. Mol Cell Biol 1994; 14:1147-52.
103. Hoang AT, Lutterbach B, Lewis BC, et al. A link between increased transforming activity of lymphoma-derived MYC mutant alleles, their defective regulation by p107, and altered phosphorylation of the c-Myc transactivation domain. Mol Cell Biol 1995; 15:4031-42.
104. Riggs KJ, Saleque S, Wong KK, et al. Yin-yang 1 activates the c-myc promoter. Mol Cell Biol 1993; 13:7487-95.
105. Shrivastava A, Saleque S, Kalpana GV, Artandi S, Goff SP, Calame K. Inhibition of transcriptional regulator yin-yang-1 by association with c-myc. Science 1993; 262:1889-92.
106. Gu W, Bhatia K, Magrath IT, Dang CV, Dalla-Favera R. Binding and suppression of the Myc transcriptional activation domain by p107. Science 1994; 264:251-4.

107. Beijersbergen RL, Hijmans EM, Zhu L, Bernards R. Interaction of c-Myc with the pRb-related protein p107 results in inhibition of c-Myc-mediated transactivation. EMBO J 1994; 13:4080-6.
108. Rustgi AK, Dysin N, Bernards R. Amino-terminal domains of c-myc and N-myc proteins mediate binding to the retinoblastoma gene product. Nature 1991; 352:541-4.
109. Lutterbach B, Hann SR. Hierarchical phosphorylation at n-terminal transformation-sensitive sites in c-myc protein is regulated by mitogens and in mitosis. Mol Cell Biol 1994; 14:5510-22.
110. Henriksson M, Bakardjiev A, Klein G, Luscher B. Phosphorylation sites mapping in the N-terminal domain of c-*myc* modulate its transforming potential. Oncogene 1993; 8:3199-209.
111. Pulverer BJ, Fisher C, Vousden K, Littlewood T, Evan G, Woodgett JR. Site-specific modulation of c-myc cotransformation by residues phosphorylated in vivo. Oncogene 1994; 9:59-70.
112. Lees E, Faha B, Dulic V, Reed S, Harlow E. Cylin E/cdk2 and cyclin A/cdk2 kinases associate with p107 and E2F in a temporally distinct manner. Genes Dev 1992; 6:1874-85.
113. Faha B, Ewen ME, Tsai LH, Livingston DM, Harlow E. Interaction between human cyclin A and adenovirus E1A-associated p107 protein. Science 1992; 255:87-90.
114. Ewen ME, Faha B, Harlow E, Livingston DM. Interaction of p107 with cyclin A independent of complex formation with viral oncoproteins. Science 1992; 255:85-7.

CHAPTER 8

Max Association with Myc

8.1 STRUCTURE AND FUNCTION

A major breakthrough occurred in the study of transcriptional regulation by c-Myc when the partner protein, Max, was identified in 1991. It had been suspected that Myc partner proteins were necessary for Myc function because Myc possesses a dimerization domain that appears necessary for cell transformation. Although early studies indicated that Myc could homodimerize, it became rapidly apparent that homodimerization of c-Myc only occurs at high protein concentrations in vitro. Blackwood and Eisenman used a functional cloning strategy to identify two alternatively spliced Max mRNAs that encode proteins of 151 and 160 amino acids.[1] The bHLHZip region of c-Myc fused to glutathione-S-transferase (GST) was used to screen a complementary DNA (cDNA) expression library. A bHLHZip protein, termed Max, was identified. Prendergast et al used a PCR strategy employing primers that matched conserved domains of Myc to identify Myn, the murine homolog of Max.[2,3] Genes or cDNAs of *max* in other species, such as Xenopus, zebrafish and chicken, have been cloned as well.[4-7] The genomic structure of the chicken gene suggests that it contains a TATA-less promoter which consists of GC-rich sequences resembling binding sites for the transcriptional factor Sp1.[4]

Mutational analysis has clearly demonstrated that the HLH Zip region of Max is required for its interaction with c-Myc. (Fig. 8.1).[1,2,8-17] Homodimerization of Max requires critical residues within the HLH domain, whereas heterodimerization of Max with c-Myc is predominantly dictated by residues found in the leucine zipper.[17] These mutational studies were accompanied by reports of hypothetical structural models of Myc and Max.[18,19] Although these models provided a period of visual excitement, they were replaced with empirically derived structures of the bHLHZip domains of Max and USF homodimers.[15,20] The crystal structures of

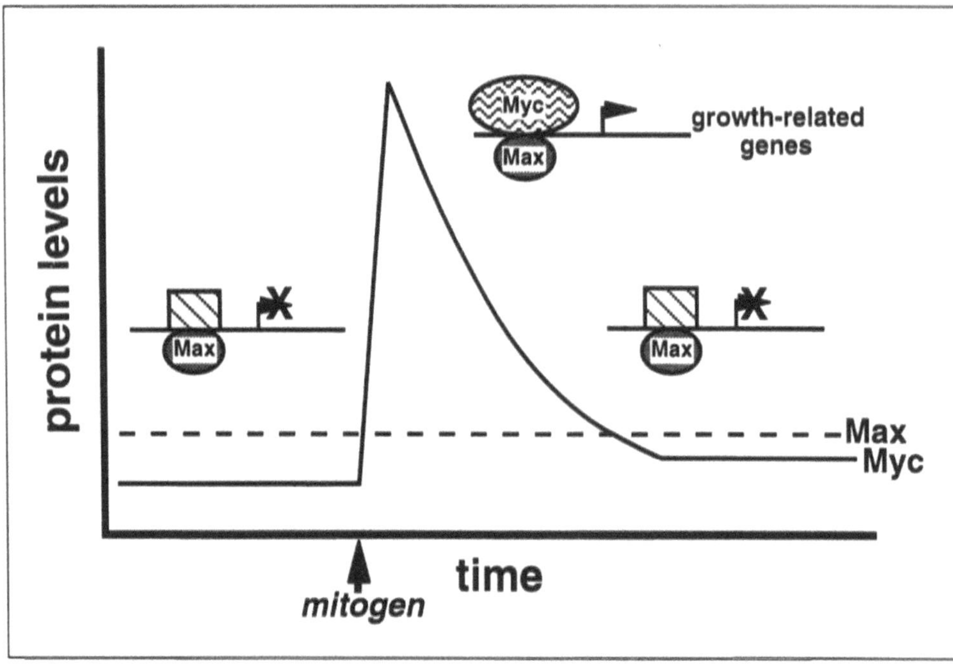

Fig. 8.1. Model of c-Myc and Max interaction. The dashed and solid lines represent relative levels of Myc and Max proteins when cells are quiescent and when stimulated to proliferate upon exposure to a mitogen. Max protein level is relatively constant while Myc protein rises sharply upon mitogenic stimulation. The change in Myc levels relative to Max accounts for the shift from inhibitory Max complexes to Myc-Max activating dimers.

these bHLHZip proteins solved in the Burley laboratory yielded a structural organization that was not predicted by either of the hypothetical models.

The gene that encodes Max has been localized to human chromosome 14q23.[21,22] As high concentrations of Max counteracts Myc activity, it was speculated that Max might be a tumor suppressor gene. Since Max is required for Myc function, however, it would be surprising to find that Max is altered in human cancers. In fact, mutations of Max in human diseases have not been reported. For example, loss of heterozygosity affecting 14q in human head and neck cancer does not involve the *max* gene (D. Sidransky, personal communication).

Max is a hydrophilic nuclear phosphoprotein that is encoded by a mRNA with a relatively stable half-life of 3 hours.[22] Alternative splicing of RNA results in at least two forms of Max protein that are expressed in cells.[23] In vivo Max is shown to be a long-lived (half-life > 24 hours) nuclear protein phosphorylated by casein kinase II.[24] The structure of Max is typical of those proteins found in the bHLHZip family in that it possesses both a basic domain by which it recognizes the sequence CANNTG, a nuclear localization signal, and a dimerization domain.[1,2,8,25] Unlike other bHLHZip family member, however, Max lacks an activation domain.[8,26] The two forms of Max that were originally isolated differ by the insertion of nine amino acids in the region amino terminal to the basic region.[1,2,8,25] Both forms are able to associate with Myc and when titrated into cells, repress transcription at high input concentrations.[25,27] In certain fibroblastic cell lines grown during logarithmic phase, the functional levels of Max appear to be limiting when compared to c-Myc levels.[28] Under such conditions, exogenous Max has been shown to activate E-box driven transcription. The DNA-binding activity of Max homodimers appears to be inhibited, at least in vitro, by amino-terminal phosphorylation of Ser2 or Ser11.[29-32] This phosphorylation, however, does not inhibit the ability of Myc-Max heterodimers to bind DNA.

Max, like c-Myc, also contains a nuclear targeting signal toward its carboxy-terminus.[8,25] Unlike c-Myc, however, a low abundancy alternatively spliced mRNA of Max has been identified.[33] This alternative *max* mRNA (termed delta *max*) contains an additional internal exon introducing an in-frame translational termination. It encodes a truncated version of Max lacking the nuclear targeting sequence. The delta-Max protein is cytoplasmic and appears to counter the effect of full length Max. In addition to the previously identified Max and delta Max proteins, sequence analysis of a 3.5-kb mRNA predicted a protein that resembles delta Max in structure.[34] Like delta Max, this protein enhanced the number of transformed foci in the *ras-myc* co-transformation assay. The existence of delta Max in cells however, has been difficult to demonstrate.

Coimmunoprecipitation and two-hybrid experiments have demonstrated that Max associates in vivo with each of the Myc family members c-Myc, L-Myc and N-Myc.[1,8,24,35] Max, however, does not dimerize with the bHLHZip transcriptional factors USF or TFE-3.[1] By means of a coimmunoprecipitation assay using anti-Myc and anti-Max antibodies, Myc and Max were found to associate in vivo and essentially all of the newly synthesized Myc was detected in a complex with Max.[24] Myc is rapidly degraded during or after its association with Max, which itself is a highly stable protein. The levels of Max expression are equivalent in quiescent, mitogen-stimulated, and cycling cells.[24,28,36] The physical association of Max with c-Myc was discovered to have functional consequences; it was observed that exogenous expression of Max augmented both c-Myc transcriptional and transformation activity.[25,28] At very high levels, exogenous Max inhibits c-Myc mediated transcriptional activation and transformation.[25,28] Dominant negative alleles of Myc proteins, which presumably interact with Max, are capable of inhibiting the function of other Myc family members.[37] These observations suggest that Max lies in the center of a network of protein-protein interactions.

Dominant negative alleles of *max* have been generated by deletion or point mutations of the basic region rendering it incapable of interacting with DNA.[2,25,38,39] These dominant negative alleles appear to counteract the function of c-Myc presumably through the formation of inactive Myc-Max heterodimers. Dominant negative *max* alleles have been shown to inhibit Myc induced phenotypes. A dominant-negative mutant called dMax was constructed and found to inhibit sequence-specific DNA binding of Myc proteins.[38] In a rat neuroblastoma model system, dMax reverts N-Myc-induced changes in cellular gene expression. A control mutant of dMax was unable to bind to N-Myc and did not revert the N-Myc-induced changes in cellular gene expression. These data support the hypothesis that N-Myc affects neuroblastoma gene expression through the formation of a DNA-binding heterodimeric complex with Max in vivo.

Dominant negative *max* interferes with differentiation of erythroleukemia cells.[39] Murine erythroleukemia (MEL) cells were transfected with a full-length wild-type (wt) human *max* gene and a basic region (bm) mutant containing a double point mutation, which abolishes specific DNA binding. Clones expressing wt-*max* grew slowly and differentiation was delayed. Clones transfected with the mutated *max* exhibited growth retardation, accumulated in the G_0/G_1 phase of the cell cycle and differentiated spontaneously. These observations suggest that large excess of wt-Max in cells enhances the formation of Max-Max growth-suppressor complexes, whereas dominant negative Max deprives the cell of growth-promoting Myc-Max heterodimers presumably by inactivating endogenous Myc and Max. It is unclear at present how Max dominant negative proteins alter the function of Myc versus those of Mad or Mxi-1 (see below) in such a way that c-Myc function is preferentially inhibited.[40-42] The requirement for the association of Myc and Max for cellular transformation has been elegantly addressed through the use of engineered Myc and Max which are constructed to contain leucine zippers that are compatible with each other but not with those of the native Myc or Max proteins.[11] These studies also demonstrate the requirement of both engineered Myc and Max for cellular transformation as well as for Myc-induced apoptosis.

A model of c-Myc and Max transcriptional regulation has been proposed based on the finding that expression of Max mRNA and protein is constitutive while that of Myc rises with cell proliferation (Fig. 8.1).[3,10,43] The model proposes that c-Myc mediated transcriptional activation of cell growth dependent genes results from a mass action shift favoring Myc-Max heterodimer formation as Myc reaches levels in excess of Max upon growth stimulation. As cells exit the cell cycle and become quiescent, Myc expression returns to basal levels resulting in a prevalence of Max inhibitory dimer formation with ensuing inhibition of transcription of growth-related genes. While this model provides a convenient working hypothesis, several studies have suggested that Max protein levels may vary with certain cellular states.[2,44]

8.2 MAX PARTNER PROTEINS

The model of transcriptional regulation by Myc and Max has proven to be too simplistic with the recent identification of at least two other Max partner proteins, Mad and Mxi-1 (Fig. 8.2).[41,45] A lambda gt11 expression library was screened with radiolabeled Max protein. One cDNA identified encodes a new member of the bHLHZip protein family, Mad, which homodimerizes poorly but binds Max in vitro, forming a specific E-box DNA binding complex.[45] Both Myc-Max and Mad-Max heterocomplexes are favored in vitro over Max homodimers. Unlike Max homodimers, the DNA binding activity of the heterodimers is unaffected by CKII phosphorylation. Mad does not associate with Myc or with other bHLH, bZip, or bHLHZip proteins. In vivo transactivation assays suggest that Mad-Max complexes oppose the transcriptional activities of Myc-Max complexes and that Max plays a central role in this network of transcription factors.

Expression of Mad varies inversely with that of Myc and increases upon differentiation of a monocyte/macrophage cell line, explaining the shift in formation of Myc-Max heterodimers to Mad-Max heterodimers 48 hours after cells are stimulated to differentiate.[40] Mad protein is induced rapidly upon differentiation of myeloid cell lines. The Mad protein is synthesized as a

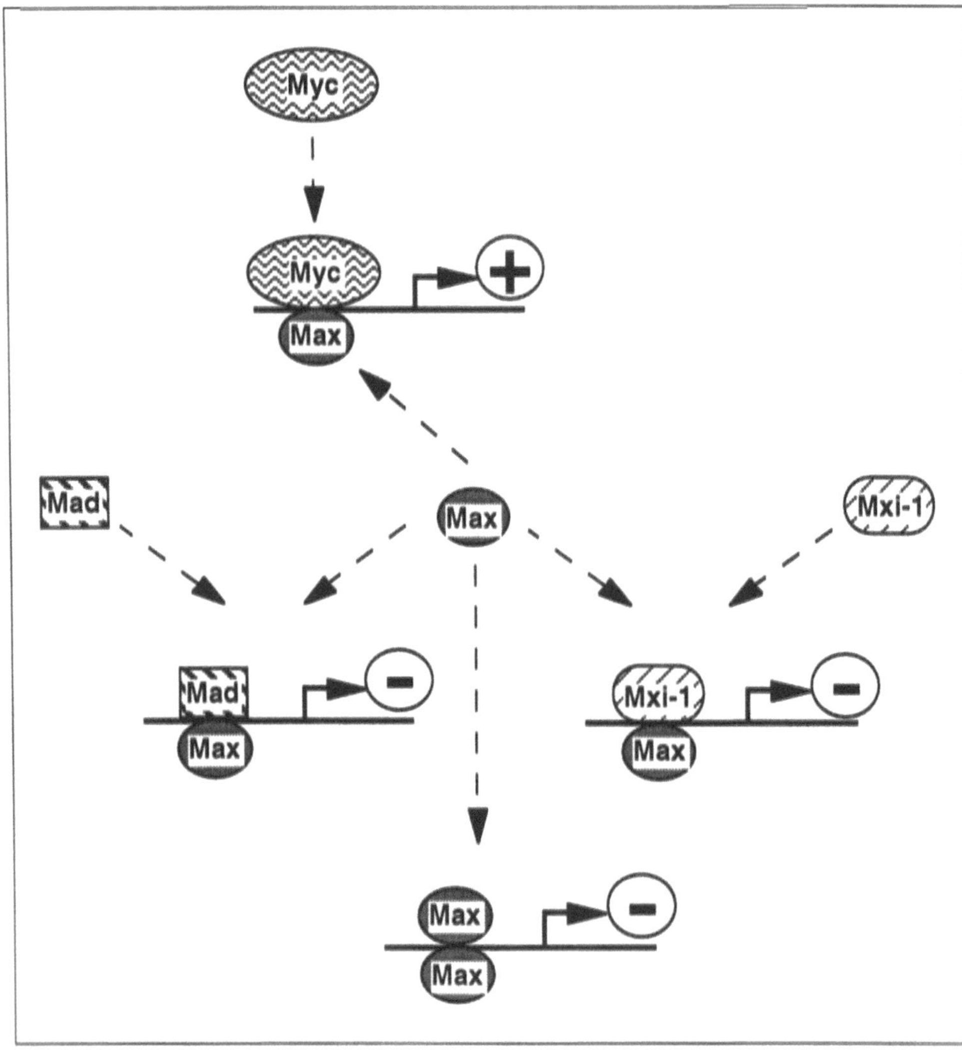

Fig. 8.2. Transcriptional regulation by Myc is likely to involve a network of proteins in vivo. Myc can function as a transcriptional activator only as a heterodimer with Max. Although no other partner proteins have been identified for Myc, at least two other partner proteins exist for Max. These proteins are Mad and Mxi-1 and their levels rise with differentiation as opposed to that of Myc. Max is also capable of forming homodimers. Max homodimers and Max-Mad and Max-Mxi-1 heterodimers bind to the E-box to inhibit c-Myc-mediated transactivation.

35-kDa nuclear phosphoprotein with an extremely short half-life ($t_{1/2}$ = 15-30 min) and can be detected in vivo in a complex with Max. Max was found complexed with Myc but not Mad in the undifferentiated U937 monocyte cell line. Mad-Max complexes, however, began to accumulate as early as 2 hr after induction of macrophage differentiation with the phorbol ester, TPA. After 48 hours of TPA treatment, only Mad-Max complexes were detectable. These results suggest that differentiation is accompanied by a switch from Myc-Max to Mad-Max heterocomplexes with differentiation. In the human myeloid cell lines HL-60 and ML-1, c-*myc* expression was downregulated after induction of differentiation, whereas the expression of *max* was minimally affected.[42] The expression of *mad* was induced and remained elevated for at least 3 days upon differentiation in both cell lines by phorbol ester (TPA), retinoic acid, vitamin D3, dimethyl sulfoxide and interferon-γ. These observations suggest that the expression of *mad* and c-*myc* is inversely regulated during induced hematopoietic differentiation.

The two-hybrid system was used to isolate another novel human protein that specifically interacts with Max. This protein, Mxi1 (for Max interactor 1), is a bHLHZip protein that interacts specifically with Max to form heterodimers that efficiently bind to the Myc-Max consensus recognition site.[41] Mxi1 does not stimulate transcription when bound to DNA via a LexA DNA binding domain in yeast. The *mxi1* mRNA is expressed in many tissues, and its expression is elevated in U-937 myeloid leukemia cells that have been stimulated to differentiate.

The expression patterns of c-Myc, Max, Mxi-1 and Mad during in vitro differentiation of the bipotent human myeloid leukemia K562 cell line suggest that specific Max heterocomplexes are associated with lineage determination.[46] K562 cells differentiate along erythroid (induced by 1-beta-D-arabinofuranosyl-cytosine) or myelomonocytic lineages (induced by 12-0-tetradecanoylphorbol-13-acetate). c-*myc* expression decreased with exposure to either differentiation inducing agent. The expression level of *max* was unchanged during myelomonocytic differentiation. In contrast, *max* mRNA and protein were downregulated during erythroid differentiation of K562 cells, suggesting that the *max* gene may be subjected to regulation during differentiation. *mxi1* mRNA expression increased during erythroid differentiation but was strongly downregulated during myelomonocytic differentiation of K562. *mad* mRNA was constitutively expressed during K562 erythroid differentiation and slightly increased during induction of the myelomonocytic pathway. These observations suggest that the Mad family of proteins may be associated with specific cellular differentiation programs.

Mad-Max and Mxi1-Max heterodimers block c-Myc mediated cell transformation.[47,48] *mad* and *mxi1* expression vectors when added to mouse c-*myc*/*ras* cotransfections dramatically reduced both the number of transformed foci and the severity of the malignant phenotype. Mad and Mxi1 specifically suppressed transformation induced by c- and N-*myc*, but not E1A. These observations support a current model for regulation of Myc function in which relative intracellular levels of Mad, Mxi1 and Myc may determine the degree of activation of Myc-related phenotypes.

The human genes for Mad and Mxi-1 have been mapped to chromosomes 2p12-13 and 10q24-25, respectively.[49-51] Since the simple model suggests that Mad and Mxi-1 counteract the function of c-Myc by both a competitive binding of Max and active transcriptional repression by Mad/Max or Mxi-1-Max, it has been proposed that these genes may be potential tumor suppressor genes. Indeed, it is intriguing to note that deletions of 10q24-25 are common in human glioblastomas and prostate cancers. Studies to date suggest that *mxi-1* displays loss of heterozygosity in about 25% of prostate cancers and up to 50% of glioblastomas (Wechsler, Dang et al, unpublished observation).[52] Mutations of *mxi-1* have been reported for human prostate cancer as well as in one case of glioblastoma (Wechsler, Dang et al, unpublished observation).[53] Whether *mxi-1* plays a role in other human malignancies remains to be established.

Searches for other Max partners by the Eisenman group have revealed a family of Mad-like proteins which includes Mxi-1 (Eisenman, personal communication). These family members share in common the Sin3 association domain and a bHLHZip domain. Another dimension to c-Myc transcriptional regulation has been added with the identification of two related mammalian cDNAs (mSin3) that encode Mad-binding proteins.[48,54] Both predicted sequences are homologous with the yeast transcription repressor Sin3, including four conserved paired amphipathic helix (PAH) domains. mSin3A and mSin3B bind to both Mad and the related protein

Mxi1. Mad-Max and mSin3 form ternary complexes that recognize the Mad-Max E box-binding site, CACGTG. The PAH2 domain of mSin3A/mSin3B and the first 25 residues of Mad, which contains a putative amphipathic alpha-helical region conserved in the family of Mad-like proteins, are required for Mad-Sin3 association. Sin3 also associates with the Mxi1 amino-terminal domain that is required for suppression of neoplastic transformation. Thus repression of c-Myc mediated transcriptional activation may occur by competition for binding between dimerization partners and by active transcriptional repression through the Mad family of proteins in association with mammalian homologs of Sin3.

Is Max the sole partner of Myc? This question remains unanswered; however, information to date suggest that Max plays a critical role in the Myc-Max network of proteins. One derivative of the neuroendocrine PC12 cell line was recently found not to contain a functional Max protein as a result of genetic alteration of the *max* gene.[55,56] Does this mean that Myc has an alternative mechanism of transforming cells? This unique PC12 cell line may provide clues to pathways that are redundant with the Myc-Max network. Experiments with knock out mice, however, suggest that Max is critical for cell growth and development, since homologous deletion of Max results in lethality very early in embryogenesis (DePinho, personal communication). Although the *max* knock out lethality suggests a critical role for *max* in embryogenesis, this finding does not rule out the possibility that other Myc partner proteins may also exist. Recently the bHLHZip domains of c-Myc were used to screen an *E. coli* based expression library to isolate an HMG-box protein, SSRP1, as a potential Myc partner. A physiological interaction between these proteins has not been demonstrated.[57]

References

1. Blackwood EM, Eisenman RN. Max: a helix-loop-helix zipper protein that forms a sequence-specific DNA-binding complex with Myc. Science 1991; 251:1211-7.
2. Prendergast GC, Lawe D, Ziff EB. Association of Myn, the murine homolog of max, with c-Myc stimulates methylation-sensitive DNA binding and ras cotransformation. Cell 1991; 65:395-407.
3. Prendergast GC, Ziff EB. A new bind for Myc. [Review]. Trends Genet 1992; 8:91-6.
4. Sollenberger KG, Kao TL, Taparowsky EJ. Structural analysis of the chicken max gene. Oncogene 1994; 9:661-4.

5. Tonissen KF, Krieg PA. Analysis of a variant max sequence expressed in xenopus laevis. Oncogene 1994; 9:33-38.
6. Schreiber-Agus N, Horner J, Torres R, Chiu FC, DePinho RA. Zebra fish myc family and max genes: differential expression and oncogenic activity throughout vertebrate evolution. Mol Cell Biol 1993; 13:2765-75.
7. King MW, Blackwood EM, Eisenman RN. Expression of two distinct homologues of Xenopus Max during early development. Cell Growth Differ 1993; 4:85-92.
8. Kato GJ, Lee WM, Chen LL, Dang CV. Max: functional domains and interaction with c-Myc. Genes Dev 1992; 6:81-92.
9. Torres R, Schreiber-Agus N, Morgenbesser SD, DePinho RA. Myc and Max: a putative transcriptional complex in search of a cellular target. [Review]. Curr Opin Cell Biol 1992; 4:468-74.
10. Blackwood EM, Kretzner L, Eisenman RN. Myc and Max function as a nucleoprotein complex. [Review]. Curr Opin Genet Dev 1992; 2:227-35.
11. Amati B, Brooks MW, Levy N, Littlewood TD, Evan GI, Land H. Oncogenic activity of the c-Myc protein requires dimerization with Max. Cell 1993; 72:233-45.
12. Crouch DH, Fisher F, Clark W, Jayaraman PS, Goding CR, Gillespie DA. Gene-regulatory properties of Myc helix-loop-helix/leucine zipper mutants: Max-dependent DNA binding and transcriptional activation in yeast correlates with transforming capacity. Oncogene 1993; 8:1849-55.
13. Davis LJ, Halazonetis TD. Both the helix-loop-helix and the leucine zipper motifs of c-Myc contribute to its dimerization specificity with Max. Oncogene 1993; 8:125-32.
14. Cole MD. Myc meets its max [Review]. Cell 1991; 65:715-716.
15. Ferre-D'Amare AR, Prendergast GC, Ziff EB, Burley SK. Recognition by Max of its cognate DNA through a dimeric b/HLH/Z domain. Nature 1993; 363:38-45.
16. Min S, Mascarenhas NT, Taparowsky EJ. Functional analysis of the carboxy-terminal transforming region of v-Myc: binding to Max is necessary, but not sufficient, for cellular transformation. Oncogene 1993; 8:2691-701.
17. Reddy CD, Dasgupta P, Saikumar P, Dudek H, Rauscher Fd, Reddy EP. Mutational analysis of Max: role of basic, helix-loop-helix/leucine zipper domains in DNA binding, dimerization and regulation of Myc-mediated transcriptional activation. Oncogene 1992; 7:2085-92.
18. Vinson CR, Garcia KC. Molecular model for DNA recognition by the family of basic-helix-loop-helix-zipper proteins. New Biol 1992; 4:396-403.
19. Halazonetis TD, Kandil AN. Predicted structural similarities of the DNA binding domains of c-Myc and endonuclease Eco RI. Science 1992; 255:464-6.
20. Ferre-D'Amare AR, Pognonec P, Roeder RG, Burley SK. Structure

and function of the b/HLH/Z domain of USF. EMBO J 1994; 13:180-9.
21. Gilladoga AD, Edelhoff S, Blackwood EM, Eisenman RN, Disteche CM. Mapping of MAX to human chromosome 14 and mouse chromosome 12 by in situ hybridization. Oncogene 1992; 7:1249-51.
22. Wagner AJ, Le Beau MM, Diaz MO, Hay N. Expression, regulation, and chromosomal localization of the Max gene. Proc Natl Acad Sci USA 1992; 89:3111-5.
23. Mäkelä TP, Koskinen PJ, Västrik I, Alitalo K. Alternative forms of Max as enhancers or suppressors of Myc-Ras cotransformation. Science 1992; 256:373-7.
24. Blackwood EM, Luscher B, Eisenman RN. Myc and Max associate in vivo. Genes Dev 1992; 6:71-80.
25. Prendergast GC, Hopewell R, Gorham BJ, Ziff EB. Biphasic effect of Max on Myc cotransformation activity and dependence on amino- and carboxy-terminal Max functions. Genes Dev 1992; 6:2429-39.
26. Min S, Taparowsky EJ. v-Myc, but not Max, possesses domains that function in both transcription activation and cellular transformation. Oncogene 1992; 7:1531-40.
27. Kretzner L, Blackwood EM, Eisenman RN. Myc and Max proteins possess distinct transcriptional activities. Nature 1992; 359:426-9.
28. Lee LA, Resar LM, Dang CV. Cell density and paradoxical transcriptional properties of c-Myc and Max in cultured mouse fibroblasts. J Clin Invest 1995; 95:900-4.
29. Berberich SJ, Cole MD. Casein kinase II inhibits the DNA-binding activity of Max homodimers but not Myc/Max heterodimers. Genes Dev 1992; 6:166-76.
30. Koskinen PJ, Vastrik I, Makela TP, Eisenman RN, Alitalo K. Max activity is affected by phosphorylation at two nh2-terminal sites. Cell Growth Differ 1994; 5:313-20.
31. Bousset K, Henriksson M, Luscherfirzlaff JM, Litchfield DW, Luscher B. Identification of casein kinase ii phosphorylation sites in max - effects on dna-binding kinetics of max homo- and myc/max heterodimers. Oncogene 1993; 8:3211-20.
32. Bousset K, Oelgeschlager MHH, Henriksson M, et al. Regulation of transcription factors c-Myc, Max, and c-Myb by casein kinase II. Cell Molec Biol Res 1995; 40:501-11.
33. Makela TP, Koskinen PJ, Vastrik I, Alitalo K. Alternative forms of Max as enhancers or suppressors of Myc-ras cotransformation. Science 1992; 256:373-7.
34. Vastrik I, Koskinen PJ, Alitalo R, Makela TP. Alternative mRNA forms and open reading frames of the max gene [published erratum appears in Oncogene 1993 Jun;8(6):1711]. Oncogene 1993; 8:503-7.

35. Wenzel A, Ciepluch C, Hamann U, Schürmann J, Schwab M. The N-Myc oncoprotein is associated in vivo with the phosphoprotein Max (p20/22) in human neuroblastoma cells. EMBO J. 1991; 10:3703-12.
36. Berberich S, Hyde-DeRuyscher N, Espenshade P, Cole M. max encodes a sequence-specific DNA-binding protein and is not regulated by serum growth factors. Oncogene 1992; 7:775-9.
37. Mukherjee B, Morgenbesser SD, DePinho RA. Myc family oncoproteins function through a common pathway to transform normal cells in culture: cross-interference by Max and trans-acting dominant mutants. Genes Dev 1992; 6:1480-92.
38. Billaud M, Isselbacher KJ, Bernards R. A dominant-negative mutant of Max that inhibits sequence-specific DNA binding by Myc proteins. Proc Natl Acad Sci USA 1993; 90:2739-43.
39. Cogliati T, Dunn BK, Bar-Ner M, Cultraro CM, Segal S. Transfected wild-type and mutant max regulate cell growth and differentiation of murine erythroleukemia cells. Oncogene 1993; 8:1263-8.
40. Ayer DE, Eisenman RN. A switch from Myc:Max to Mad:Max heterocomplexes accompanies monocyte/macrophage differentiation. Genes Dev 1993; 7:2110-9.
41. Zervos AS, Gyuris J, Brent R. Mxi1, a protein that specifically interacts with Max to bind Myc-Max recognition sites [published erratum appears in Cell 1994 Oct 21;79(2):following 388]. Cell 1993; 72:223-32.
42. Larsson LG, Pettersson M, Oberg F, Nilsson K, Luscher B. Expression of mad, mxi1, max and c-myc during induced differentiation of hematopoietic cells: opposite regulation of mad and c-myc. Oncogene 1994; 9:1247-52.
43. Kato GJ, Dang CV. Function of the c-Myc oncoprotein. [Review]. FASEB J 1992; 6:3065-72.
44. Dunn BK, Cogliati T, Cultraro CM, Barner M, Segal S. Regulation of murine max (myn) parallels the regulation of c-myc in differentiating murine erythroleukemia cells. Cell Growth Differ 1994; 5:847-854.
45. Ayer DE, Kretzner L, Eisenman RN. Mad: a heterodimeric partner for Max that antagonizes Myc transcriptional activity. Cell 1993; 72:211-22.
46. Delgado MD, Lerga A, Canelles M, Gomezcasares MT, Leon J. Differential regulation of max and role of c-myc during erythroid and myelomonocytic differentiation of K562 cells. Oncogene 1995; 10:1659-65.
47. Lahoz EG, Xu L, Schreiber-Agus N, DePinho RA. Suppression of Myc, but not E1a, transformation activity by Max-associated proteins, Mad and Mxi1. Proc Natl Acad Sci USA 1994; 91:5503-7.
48. Schreiber-Agus N, Chin L, Chen K, et al. An amino-terminal domain of Mxi1 mediates anti-Myc oncogenic activity and interacts with a homolog of the yeast transcriptional repressor SIN3. Cell 1995; 80:777-86.

49. Edelhoff S, Ayer DE, Zervos AS, et al. Mapping of two genes encoding members of a distinct subfamily of MAX interacting proteins: MAD to human chromosome 2 and mouse chromosome 6, and MXI1 to human chromosome 10 and mouse chromosome 19. Oncogene 1994; 9:665-8.
50. Shapiro DN, Valentine V, Eagle L, Yin X, Morris SW, Prochownik EV. Assignment of the human MAD and MXI1 genes to chromosomes 2p12-p13 and 10q24-q25. Genomics 1994; 23:282-5.
51. Wechsler DS, Hawkins AL, Li X, Jabs EW, Griffin CA, Dang CV. Localization of the human Mxi1 transcription factor gene (MXI1) to chromosome 10q24-q25. Genomics 1994; 21:669-72.
52. Albarosa R, Didonato S, Finocchiaro G. Redefinition of the coding sequence of the MXI1 gene and identification of a polymorphic repeat in the 3' non-coding region that allows the detection of loss of heterozygosity of chromosome 10q25 in glioblastomas. Human Genet 1995; 95:709-11.
53. Eagle LR, Yin XY, Brothman AR, Williams BJ, Atkin NB, Prochownik EV. Mutation of the MXI1 gene in prostate cancer. Nature Gen 1995; 9:249-55.
54. Ayer DE, Lawrence QA, Eisenman RN. Mad-Max transcriptional repression is mediated by ternary complex formation with mammalian homologs of yeast repressor Sin3. Cell 1995; 80:767-76.
55. Ribon V, Leff T, Saltiel AR. C-myc does not require max for transcriptional activity in pc-12 cells. Mol Cell Neurosci 1994; 5:277-82.
56. Hopewell R, Ziff EB. The nerve growth factor-responsive PC12 cell line does not express the Myc dimerization partner Max. Mol Cell Biol 1995; 15:3470-8.
57. Bunker CA, Kingston RE. Identification of a cDNA for SSRP1, an HMG-box protein, by interaction with the c-Myc oncoprotein in a novel bacterial expression screen. Nucleic Acids Res 1995; 23:269-76.

CHAPTER 9

DNA Binding Properties of Myc

Recent studies have finally substantiated the long-held hypothesis that c-Myc binds to specific DNA sequences. Myc binds the palindromic consensus sequence CANNTG that is recognized by all basic regions of members of the bHLHZip family.[1] Studies using pools of random oligonucleotides, the polymerase chain reaction, and recombinant fusion Myc protein along with electrophoretic mobility gel shift assay indicate that the preferred core binding site for Myc is CACGTG.[1] Full length c-Myc is unable to bind DNA by itself, unless a significant portion of its N-terminal is removed.[2] Although truncated c-Myc binds the consensus sequence as a homodimer in vitro only when present at very high protein concentrations, full-length c-Myc binds DNA more efficiently as a heterodimer with its partner protein Max.[1-3] The core sequence CACATG was also bound by Myc-Myc, Myc-Max, and Max-Max to similar extent.[2,4-8] Additional studies indicate that the binding of Myc to this core hexanucleotide shows preference for specific flanking nucleotides as well: GAC(cacgtg)GTC.[9,10] The flanking sequences may be important in Myc DNA binding as another bHLHZip protein, USF, recognizes the same canonical sequence, but prefers alternative flanking nucleotides.[11,12]

The specificity of Myc function may be dictated by its ability to bind unique DNA sites. Non-canonical Myc binding sites have been identified from pools of random oligonucleotides.[12] Of specific interest is the sequence CGCATG which appears to be bound by Myc and Max but not by the related bHLHZip transcription factor USF. In addition, a separate study suggests that c-Myc is able to bind a C/EBP-like DNA binding site and transactivate through this sequence.[13] Interestingly, only the AUG-initiated protein (62kDa) but not the CUG-initiated form (64kDa) was capable of transactivation. The physiological significance of these observations remains unclear, and thus further studies are necessary to determine the significance of $_{p}62$ and $_{p}64$ c-Myc proteins. More recently, in vitro data suggest that Myc in complex with the general transcription factor TFII-I can bind initiator sequences with a consensus of YAYTCYYY (Y = pyrimidine).[14,15] While the identification of these specific DNA binding sequences for c-Myc indicate an ability of c-Myc to bind specific DNA sequences, the physiological relevance of these binding specificities is only beginning to emerge.

The HLHZip interface promotes dimerization that aligns the basic regions of Myc and Max in a heterodimeric complex in such a way as to allow critical amino acid residues to contact specific DNA sequences.[16,17] The specificity of recognition by the c-Myc basic domain for this core consensus sequence is partly held by a single amino acid in the basic region, Arg^{367}. bHLHZip proteins are assigned to either class A or class B proteins according to their cognate core DNA binding sites.[18] All proteins of class B, which c-Myc and Max are members of, recognize the sequence CACGTG as opposed to CAGCTG bound by proteins in class A, such as MyoD, E12 and E47. All of the proteins in class B contain the critical arginine at position 367. Protein crystallography revealed that this residue directly contacts the CG dinucleotide pair of the consensus sequence as it lies in the major groove of DNA, confirming its important role in specific DNA binding. The molecular basis for flanking sequence (flanking CANNTG) preference by the Myc family of proteins has not been revealed by current crystallographic data. It has been hypothesized that loop sequences of the helix-loop-helix region may be responsible for flanking site preference.[16,19,20]

A major paradox concerning c-Myc DNA binding ability remains unresolved. Although purified recombinant or in vitro translated c-Myc and Max are capable of tightly binding DNA in vitro as determined by gel mobility shift assays, many investigators have had difficulties detecting Myc-Max complexes from nuclear extracts. In contrast, a related transcription factor, USF, that binds the same core sequence as does Myc is abundant in nuclear extract. In one study, a minor shifted band (bound to a labeled CACGTG probe) detected by using HeLa nuclear extract in a gel mobility shift assay was altered by Myc or Max specific antibodies.[21] It is possible that c-Myc is an inherently biochemically labile protein; however this issue remains to be studied. Max, on the other hand, is both metabolically and biochemically stable.

Since higher order organization of nucleoprotein complexes is likely to be relevant in vivo and DNA binding proteins are able to distort the structure of DNA to which they are bound, the ability of Myc and Max to bend DNA was examined.[22] These studies were performed using gel mobility shift assays with sets of special radiolabeled DNA probes that allow determination of DNA bending. Similar to Fos and Jun, which bend DNA in opposite orientations,[23,24] Max homodimers bend DNA away from the protein whereas truncated Myc homodimers bend DNA toward the protein (Fig. 9.1). Myc and Max heterodimers appear to bend DNA similarly to Max homodimers. Caveats in the interpretation of these studies include potential differences in the strength of protein-protein interactions as well as protein-nucleic acid interactions. It is clear, however, that Myc and Max behave differently in these assays. It is notable that the crystal structures for Max and USF did not reveal bending of a short oligonucleotide constrained within a crystal lattice.[16,17]

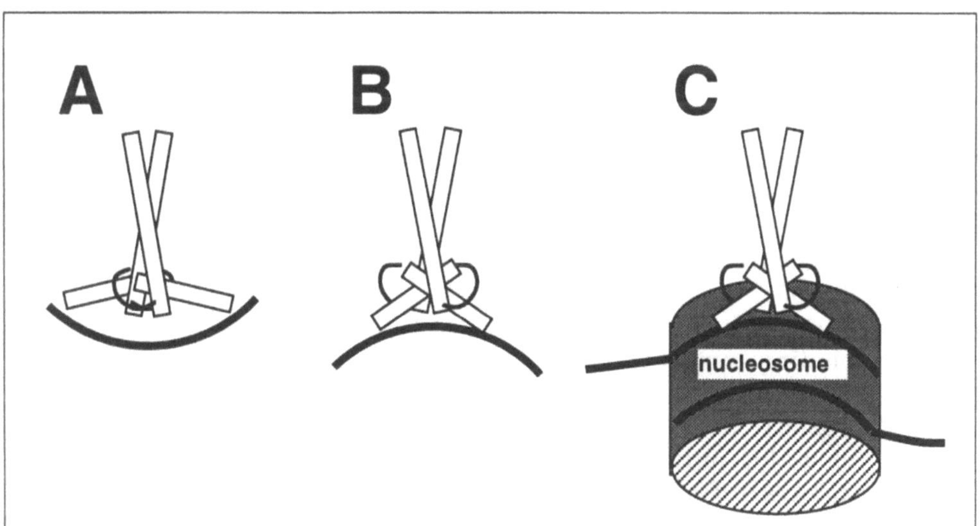

Fig. 9.1. DNA bending by Myc and Max proteins. A. Truncated Myc homodimers bend DNA toward the protein. B. Max homodimers bend DNA away from the protein. C. Myc-Max heterodimers bend DNA away from the protein like Max homodimers. Such DNA bending may be significant in the the ability of transcription factors to bind nucleosomal DNA.

To begin to understand the significance of these observations, the ability of these dimers to bind nucleosomal DNA was studied since it is becoming widely accepted that nucleosomes regulate transcription.[25-31] Using synthetic nucleosomes containing an E-box sequence, it was observed that Max homodimers efficiently bound nucleosomes, whereas artificial Myc homodimers inefficiently bound an E-box sequence wrapped around core histones in nucleosomes.[32] Myc-Max heterodimers also bound nucleosomes; however the affinity of the heterodimers for nucleosomal DNA appears to be significantly less than that of Max homodimers. Whilst these observations are intriguing from a biophysical and structural viewpoint, their physiological significance remains unestablished.

References

1. Blackwell TK, Kretzner L, Blackwood EM, Eisenman RN, Weintraub H. Sequence-specific DNA binding by the c-Myc protein. Science 1990; 250:1149-51.
2. Kato GJ, Lee WM, Chen LL, Dang CV. Max: functional domains and interaction with c-Myc. Genes Dev 1992; 6:81-92.
3. Blackwood EM, Luscher B, Eisenman RN. Myc and Max associate in vivo. Genes Dev 1992; 6:71-80.
4. Prendergast GC, Ziff EB. Methylation-sensitive sequence-specific DNA binding by the c-Myc basic region. Science 1991; 251:186-9.
5. Solomon D, Amati B, Land H. Distinct DNA binding preferences for the c-myc:max and max:max dimers. Nucleic Acids Res 1993; 21:5372-6.
6. Berberich S, Hyde-DeRuyscher N, Espenshade P, Cole M. *Max* encodes a sequence-specific DNA-binding protein and is not regulated by serum growth factors. Oncogene 1992; 7:775-9.
7. Prochownik EV, VanAntwerp ME. Differential patterns of DNA binding by Myc and Max proteins. Proc Natl Acad Sci USA 1993; 90:960-4.
8. Kerkhoff E, Bister K, Klempnauer KH. Sequence-specific DNA binding by Myc proteins. Proc Natl Acad Sci USA 1991; 88:4323-7.
9. Papoulas O, Williams NG, Kingston RE. DNA Binding Activities of c-Myc purified from eukaryotic cells. J Biol Chem 1992; 267:10470-80.
10. Halazonetis TD, Kandil AN. Determination of the c-Myc DNA-binding site. Proc Natl Acad Sci USA 1991; 88:6162-6.
11. Bendall AJ, Molloy PL. Base preferences for DNA binding by the bHLH-Zip protein USF: effects of MgCl2 on specificity and comparison with binding of Myc family members. Nucleic Acids Res 1994; 22:2801-10.

12. Blackwell TK, Huang J, Ma A, et al. Binding of myc proteins to canonical and noncanonical DNA sequences. Mol Cell Biol 1993; 13:5216-24.
13. Hann SR, Dixit M, Sears RC, Sealy L. The alternatively initiated c-Myc proteins differentially regulate transcription through a noncanonical DNA-binding site. Genes Dev 1994; 8:2441-52.
14. Roy AL, Carruthers C, Gutjahr T, Roeder RG. Direct role for Myc in transcription initiation mediated by interactions with TFII-I. Nature 1993; 365:359-61.
15. Roy AL, Meisterernst M, Pogenec P, Roeder RG. Cooperative interaction of an initiator-binding transcription initiation factor and the helix-loop-helix activator USF. Nature 1991; 354:245-8.
16. Ferre-D'Amare AR, Prendergast GC, Ziff EB, Burley SK. Recognition by Max of its cognate DNA through a dimeric b/HLH/Z domain. Nature 1993; 363:38-45.
17. Ferre-D'Amare AR, Pognonec P, Roeder RG, Burley SK. Structure and function of the b/HLH/Z domain of USF. EMBO J 1994; 13:180-9.
18. Dang CV, Dolde C, Gillison ML, Kato GJ. Discrimination between related DNA sites by a single amino acid residue of Myc-related basic-helix-loop-helix proteins. Proc Natl Acad Sci USA 1992; 89:599-602.
19. Halazonetis TD, Kandil AN. Predicted structural similarities of the DNA binding domains of c-Myc and endonuclease Eco RI. Science 1992; 255:464-6.
20. Davis LJ, Halazonetis TD. Both the helix-loop-helix and the leucine zipper motifs of c-Myc contribute to its dimerization specificity with Max. Oncogene 1993; 8:125-32.
21. Littlewood TD, Amati B, Land H, Evan GI. Max and c-Myc/Max DNA-binding activities in cell extracts. Oncogene 1992; 7:1783-92.
22. Wechsler DS, Dang CV. Opposite orientations of DNA bending by c-Myc and Max. Proc Natl Acad Sci USA 1992; 89:7635-7639.
23. Kerppola TK, Curran T. Fos-Jun heterodimers and Jun homodimers bend DNA in opposite orientations: Implications for transcription cooperativity. Cell 1991; 66:317-26.
24. Kerppola TK, Curran T. DNA bending by Fos and Jun: the flexible hinge model. Science 1991;254:1210-4.
25. Adams CC, Workman JL. Nucleosome displacement in transcription. Cell 1993; 72:305-8.
26. Clark DJ, Felsenfeld G. A nucleosome core is transferred out of the path of a transcribing polymerase. Cell 1992; 71:11-22.
27. van Holde K. The omnipotent nucleosome. Nature 1993; 362:111-12.
28. Grunstein M. Nucleosomes: Regulators of Transcription. Trends Genet. 1990; 6:395-400.

29. Wolffe AP. New insights into chromatin function in transcriptional control. FASEB J. 1992; 6:3354-61.
30. Hayes JJ, Wolffe AP. The Interaction of Transcription Factors with Nucleosomal DNA. BioEssays 1992; 14:597-603.
31. Felsenfeld G. Chromatin as an essential part of the transcriptional mechanism. Nature 1992; 355:219-24.
32. Wechsler DS, Papoulas O, Dang CV, Kingston RE. Differential binding of c-myc and max to nucleosomal DNA. Mol Cell Biol 1994; 14:4097-107.

CHAPTER 10

Myc Target Genes in Cell Proliferation and Programmed Cell Death

In the previous chapters, we reviewed the dissection of the c-Myc molecule and its structural organization that typifies a transcription factor. c-Myc function is regulated by a complicated network of proteins that developed through millions of years of evolution and is likely to include a large repertoire of interacting proteins. Yet, the mechanism by which c-Myc or its retroviral counterpart v-Myc transforms cells remains largely unknown. Clues to the molecular basis of c-Myc mediated cellular transformation may be revealed by studies that attempt to identify target genes and events linking the deregulated expression of Myc and transformed phenotypes.

10.1 PUTATIVE Myc TARGET GENES

To provide a common ground for discussion, we must define the meaning of target genes. In the broadest sense, any gene that c-Myc 'transactivates' may be considered a target. This definition, however, does not distinguish between direct and indirect targets and does not address the relevance of the targets to Myc-mediated phenotypes. A direct target gene is defined as a gene whose expression is altered as a consequence of a direct interaction of the c-Myc protein with the gene regulatory *cis*-acting DNA elements or with the *trans*-acting factors that bind these *cis* elements. Thus, the induction of an artificially construed promoter-reporter plasmid is insufficient to establish the identity of a target gene. Regulation of the endogenous cellular target gene expression by c-Myc is also required to establish the authenticity of a target gene. Conversely, the absence of transactivation of artificial promoter-reporter constructs by c-Myc should not be sufficient to rule out a target gene that may require chromosomal context as well as other *trans*-acting factors to be transactivated.

The time course of induction of a direct target gene presumably follows the expression of Myc within several hours. This view, however, may unnecessarily preclude the possibility of genes whose expression depends on the presence of c-Myc protein but whose expression is delayed and requires other *trans*-acting factors that appear late after the induction of Myc expression. Of course, this would be possible providing that the half-life of DNA-bound c-Myc protein is extended. Indirect target genes of c-Myc may be considered as those whose expressions are altered as a consequence of the expression of the direct Myc target genes and whose expressions are directly connected to c-Myc mediated phenotypes such as cellular proliferation, transformation or apoptosis. Although the search for target genes usually implies identification of the direct ones, it has become clear that indirect targets may provide the much needed clues to link deregulated c-Myc expression to neoplastic transformation.

The criteria to identify relevant target genes have not been formally agreed upon. The lack of a clear set of criteria stems in part from difficulties with providing evidence that a putative target gene is biologically relevant. This difficulty is further confounded by the fact that target genes may be necessary but not sufficient for Myc-associated phenotypes. In such case, antisense experiments appear to be a reasonable approach for establishing the necessity of a specific target gene for a Myc-associated phenotype. The difficulty with this experimental approach, however, is when a target gene (for example a cyclin gene) is one whose disruption leads to dramatic cellular changes. That is, if cyclin E, for example, is a Myc target, then disruption of cyclin E should lead to growth arrest since abrogation of Myc activity correlates with growth arrest. The fallacy of this argument is that disruption of cyclin E expression in itself leads to growth arrest and hence it could be a target of any growth promoting transcription factor including E2F. In light of these caveats, we will review putative c-Myc target genes and attempt to use the above definitions to guide our discussion.

10.1.1 IDENTIFYING TARGET GENES THROUGH DIFFERENTIAL OR SUBTRACTIVE CLONING

Approaches to identifying target genes vary in different studies. In general, there are two major approaches. One is to guess a putative target gene based on the biology of c-Myc. The other approach is to identify genes that are differentially expressed as a function of enforced Myc expression. Early studies rely on the latter approach since our knowledge of cell cycle regulation only blossomed recently allowing for better guesswork. Putative Myc target genes are shown in Figure 10.1.

Using differential cloning, early studies by the Cole group suggested that plasminogen activator inhibitor-1 (PAI-1) gene expression is augmented by c-Myc.[1,2] In fact, additional studies suggest that c-Myc alters the levels of PAI-1 mRNA at the post-transcriptional level. A recent report also suggests that Myc plays a role in post-transcriptional regulation of rRNA levels; whether this is a direct effect of c-Myc was not addressed however.[3] Although PAI-1 may be a target of c-Myc (it is notable that the PAI-1 gene promoter contains several E-boxes), it is not clear what the role of PAI-1 is in the biology of c-Myc. TGFβ1 induces PAI-1 in cultured fibroblasts. The TGFβ1 responsive element was mapped to a sequence located about 550 bp upstream of the human gene start site.[4] It is composed of overlapping sites for CCAAT-binding protein (NF1/CTF1) and the E-box sequence, CACGTG. The upstream stimulatory factor or USF was found to bind the E-box as determined by supershift experiment with an anti-USF antibody. c-Myc does not appear to participate in this pathway.

Differential cloning by the Freytag group demonstrated that collagen gene expression (types α1(I), α2(I), and α3(VI)) is down-regulated in fibroblasts over-expressing c-Myc.[5,6] This group went on further to demonstrate that c-Myc suppresses promoters driven by NF1 DNA binding sites.[6] In addition, the form of transcription factor NF1, which exists as a family of polypeptides, was altered by enforced c-Myc expression. These forms appear to be phosphorylated NF1 and are found in proliferating cells. NF1 is required for the utilization of the collagen α2(I) promoter. In this case, the promoters of collagen genes do not appear to contain

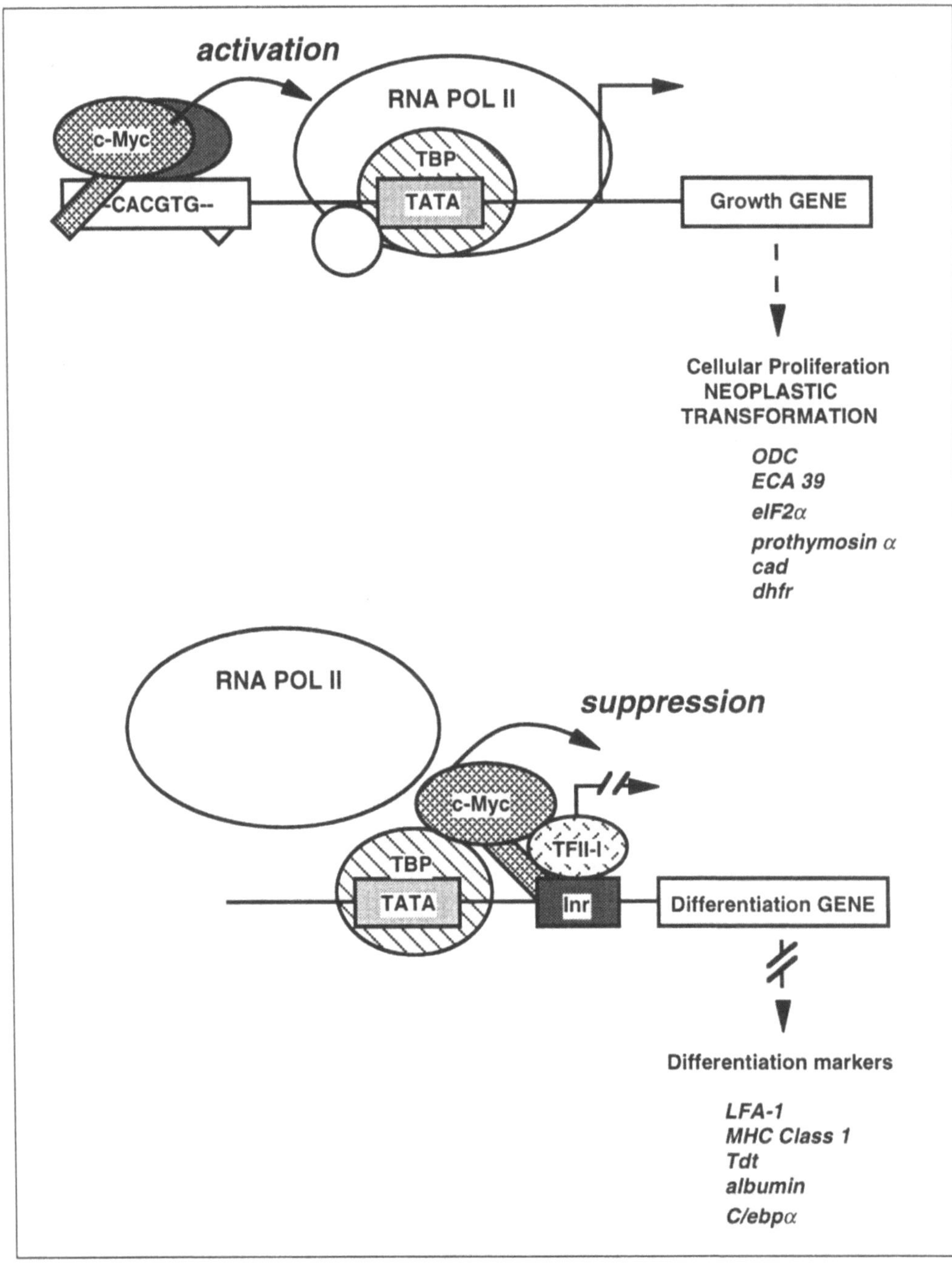

Fig. 10.1. Schematic diagram of the putative interaction between Myc and the E-box and Myc and the initiator element. Myc transactivates genes through the E-box. This is the presumed mechanism by which Myc transactivates putative growth-related genes, such as ornithine decarboxylase (ODC), ECA 39, and eIF 2α (top diagram). Myc transrepression occurs through the initiator element, designated Inr. Transrepression may involve the general transcription factor, TFII-I, which was shown to interact directly with both Myc and the Inr in vitro (lower diagram). Genes such as LFA-1, MHC class I, CEBPα, and albumin, have been reported to be down-regulated by c-Myc and contain an Inr. RNA pol II = RNA polymerase II; TBP = TATA-binding protein.

direct binding sites for c-Myc; nevertheless, c-Myc reproducibly downregulates collagen expression in fibroblasts. Thus, it appears that collagen is an indirect target of c-Myc.[6] Myc does not appear to be the only transforming gene that downregulates collagen gene expression. Cells transformed by other oncogenes, such as p53, *mos*, *ras*, *raf*, or *src* also resulted in decrease collagen gene expression.[7-12] It is intriguing to note that ectopic expression of NF1 has been shown in avian cells to inhibit the transforming activity of v-Myc as well as a variety of other oncogenes.[13] Moreover, ectopic expression of collagen has been shown to inhibit anchorage-independent growth of a chemically transformed cell line.[14] Fusion of normal cells with benzopyrene-treated Syrian hamster embryo cells augments collagen gene expression and suppresses the transformed phenotype.[15] Based on these observations, it may be speculated that c-Myc suppresses transcription of certain genes through alteration of NF1 activity. Moreover, indirect targets such as collagen genes may play a key role in Myc-associated phenotype such as anchorage-independent growth.

An alternative approach to identifying c-Myc target genes is to clone cDNAs corresponding to differentially expressed mRNAs in a system in which c-Myc expression is inducible. Using a Myc-estrogen receptor (ER) chimeric system, Eilers and Bishop showed that exposure of Rat 1a fibroblasts expressing the MycER fusion to estradiol causes the cells to enter S-phase in the absence of serum.[16,17] This system was exploited to identify Myc targets. A differentially expressed gene encoding alpha-prothymosin was identified.[17] Alpha-prothymosin protein levels have been shown to correlate with cell proliferation, although its function remains unknown.[18-20] Eilers and coworkers reported that E-boxes located in intron 1 of the prothymosin gene mediates transactivation of this gene by c-Myc.[21,22] It should be noted that the induction of reporter constructs bearing these sequences was only about 2- to 3-fold. Subsequently, this group reported the existence of AP-2 sites overlapping E-boxes in the prothymosin gene as well as other putative c-Myc target genes.[23] AP-2 appears to suppress transcription in this context, although the effects of AP-2 on c-Myc mediated phenotypes have not been demonstrated. A separate study from the Berger laboratory, however, was unable to demonstrate an induction of prothymosin gene expression or endogenous prothymosin protein levels by c-Myc in transfected cells (S. Berger, unpublished data). Further, the induction of prothymosin by c-Myc appears to be cell type specific, since c-Myc does not induce prothymosin in 3T3 fibroblasts.[21,24] The fact that prothymosin gene expression is cell specific does not exclude it as a c-Myc target. Rather, it is possible that the genetic background of a cell determines which targets are rate-limiting for cell proliferation. On the other hand, cell type specific effects suggest that prothymosin is not a rate limiting target in all cells and probably does not mediate Myc-induced neoplastic transformation.

A unique effort has been made to clone cDNAs corresponding to mid-G_1 serum-response genes which are also upregulated in two inducible c-Myc cell lines.[24] This work by Tavtigian[24] in the Wold laboratory showed that transcription of the lactate dehydrogenase and ornithine decarboxylase genes are upregulated by c-Myc. A cDNA sequence with limited homology to methylenetetrahydrofolate synthetase (clone I-8-29) sequence was also upregulated. Alpha-prothymosin was found to be upregulated only in certain cells. p53 appears to be upregulated by c-Myc but only in the zinc inducible Myc NIH3T3 cell line. Among genes that may be repressed by c-Myc, alpha-1(III) collagen was repressed in the NIH3T3 cell line but remained unchanged in the MycER Balb/c 3T3 cells after exposure to estrogen. This work is clearly the most comprehensive study to identify c-Myc target genes using several models in parallel. It extends the observations that certain targets are likely to be authentic such as the ornithine decarboxylase gene. This study also points out the limitations in studying c-Myc target genes in only selected cell lines since there may be cell specific effects of c-Myc.

Yet another study utilizing subtraction cloning identified an embryonically expressed gene, ECA39, as a Myc target that appears to have a cell specific response to c-Myc.[25] In this study a cDNA library, which was generated from brain tumor caused by the HTLV-1 LTR-*myc* transgene, was depleted of cDNAs found in normal brain through subtractive hybridization.[26] One clone that was highly expressed in Myc-induced brain tumors and in Myc-mediated T-lymphomas in transgenic animals was selected for further study and found to be a previously cloned cDNA termed ECA39. This gene is not overexpressed, however, in other transgenic c-Myc induced tumors such as mammary tumor or B-cell lymphomas. Thus ECA39 appears to be associated with Myc expression only in specific types of tumors derived from transgenic animals. The promoter region of ECA39 revealed a CACGTG E-box 105 bp 3' of the transcriptional start site. Cotransfection experiments with the ECA39 promoter-reporter construct and c-Myc expression vector suggest that c-Myc is capable of activating transcription through this promoter region in an E-box dependent manner. Furthermore, the authors show gel mobility shift experiments suggesting that the E-box element is shifted by nuclear extract from brain tumors. The shifted band demonstrate an altered mobility in the presence of a c-Myc monoclonal antibody. These observations are interpreted to suggest that ECA39 is a direct target of c-Myc. It is curious that nuclear extracts from brain tumors did not contain the transcription factor USF that is ubiquitous, abundant and known to bind to the E-box core CACGTG. The evidence presented suggest that ECA39 may be a target of c-Myc in certain tumors, but its expression is not altered in other Myc-associated tumors. Further studies are necessary to establish a biological link between c-Myc, ECA39 and neoplastic transformation. The function of ECA39 remains as yet unknown.

10.2 CANDIDATE Myc TARGET GENES AND CELL PROLIFERATION

An alternative to a search for differentially expressed genes is to test for target genes based on the biology of c-Myc. The ornithine decarboxylase (ODC) gene represents such a target. The expression of ODC was noted to correlate with the level of c-Myc expression.[27] ODC has been known to be a key enzyme in the synthesis of polyamines which are required for cell proliferation. In fact, deregulated expression of ODC appears to contribute to neoplastic transformation and an inhibitor of ODC, difluoromethylornithine (DFMO), inhibits cell growth in vitro and has been used as an anti-neoplastic agent in clinical trials.[28] ODC has now been shown to be a target for c-Myc in several independent studies.[29-31] Specifically, the time course of ODC is compatible with it being a Myc target. The ODC upstream (-491 bp to -474 bp in the human ODC promoter) sequence contains an E-box that has been implicated in Myc regulation.[30,32] In other studies, however, two E-boxes in intron 1 have been shown to respond to c-Myc.[29] The occurrence of intronic E-boxes in both prothymosin and ODC genes and an E-box element downstream of the start site of ECA39 suggest that the position of E-boxes may be important for c-Myc mediated transactivation. Subsequent work by the Cleveland group failed to demonstrate significant differences in transactivation when the E-box elements were moved to various positions relative to the transcriptional start site.[33]

Studies of gene promoters have led to the recognition of the CACGTG E-box sequence in a variety of genes. It should be noted that this E-box can be bound by USF, TFE-3 or TFE-B in addition to c-Myc.[34-36] So its existence in promoter regions should include the possibility that USF, TFE-3 or TFE-B are the transactivators. In fact, it is as yet unclear what the role is of the ubiquitous USF in the regulation of these so-called Myc E-boxes. Nonetheless, several promoters have been proposed to be c-Myc targets based on this criterion. The promoters of genes for dihydrofolate reductase (*dhfr*) and carbamoyl-phosphate synthase (*cad*) contain E2F as well as the Myc E-box sequences.[37,38] Both genes are proposed to be c-Myc target genes based on the time course of expression of these genes relative to c-*myc* expression, although *dhfr* expression may be dependent on E2F and is not inhibited by dominant negative c-*myc* alleles.[38] In the case of *dhfr*, electrophoretic mobility shift assays with the putative Myc binding sequence revealed bands that were supershifted with anti-c-Myc antibodies.

Expression of the *cad* gene increases at the G_1/S-phase boundary. Miltenberger et al have mapped the growth-dependent response element in the hamster *cad* gene to a palindromic E-box sequence, CCACGTGG, which is 65 bp 3' of the initiation site and immediately 5' of an E2F binding site.[38] Growth response was determined by the serum stimulation response of promoter-reporter constructs. Mutations directed at the E-box sequences caused a dramatic decrease in growth-response of the promoter as compared to mutations directed at the E2F binding site. Nuclear proteins that bound to the E-box sequence in a cell cycle dependent manner were blocked in electrophoretic mobility shift assays with anti-USF and anti-Max antibodies. Specific c-Myc gel shift activities were not detected; however, dominant negative alleles of c-Myc were able to block *cad* gene expression but not the expression of *dhfr*. These observations suggest that c-Myc-Max heterodimers may play a pivotal role in the expression of the *cad* gene at the G_1/S-phase boundary, although the participation of *cad* in Myc-mediated phenotypes has not been studied.

The emergence of cyclins, CDKs and CDIs as cell cycle regulators has provided a novel view of c-Myc function. Overexpression of c-Myc has been known to shorten the G_1 phase of the cell cycle of fibroblasts.[39] Induction of c-Myc activity in the MycER fusion system or with zinc-inducible Myc is tightly associated with entry into S-phase.[16,17] Removal of a copy of *c-myc* through homologous recombination results in Rat 1 fibroblasts that have a delayed entry into S-phase and a prolonged G_1 phase.[40-42] These observations suggest a connection between c-Myc and the cell cycle machinery.

Jansen-Durr et al first demonstrated a connection between deregulated c-Myc expression in BALB/c 3T3 fibroblasts and increased cyclin A and E mRNA levels.[43] Estrogen activated MycER suppresses the expression of cyclin D1 very early in the cell cycle whereas it induces the expression of cyclins A and E.[44] Another study contrasts with that of Jansen-Durr et al; cyclin D1 expression was shown to be induced in the activated MycER system.[45] Cyclin D1 protein level was found to be elevated, although its mRNA level remained essentially unchanged, in yet another MycER system.[46] It is postulated in this study that cyclin D1 protein level was augmented by elevated levels of the translational initiation factor eIF4E that preferentially enhances the translation of cyclin D_1 mRNA (see below). The conflicting data regarding the potential regulation of cyclin D1 by c-Myc is presently unresolved.

The correlation of c-Myc expression with elevated cyclin A expression has been independently observed in several studies.[42,43,45,47] Using heterozygous *myc* cells derived by homologous recombination, it was observed that reduced *c-myc* expression is associated with a delayed expression of cyclin E and reduced expression of cyclin A protein.[42] Overexpression of c-Myc correlated with an increase in cyclin D_1 protein level, early expression of cyclin E and an increase in cyclin A protein levels. Cdk2 protein levels were unaltered by changes in *c-myc* expression. The effect of c-Myc on cyclin D_1 expression is at present unclear, since the results of different studies are conflicting. The observation that c-Myc expression correlates with that of cyclin A has been independently verified by all independent studies published to date. The available sequence of the cyclin A promoter (>3kb) does not contain a consensus c-Myc E-box.[48,49] The critical element in the cyclin A promoter that regulates its expression appears to be a binding site for ATF/CREB.[50-52] It is as yet unclear whether c-Myc directly activates cyclin A (which is doubtful when the time course of appearance of cyclin A after c-Myc induction is considered) or that c-Myc may modulate other factors such as ATF to alter cyclin A expression.

Induction of cyclin A and E expression may in part explain a number of c-Myc-related phenotypes. As already discussed, deregulated expression of c-Myc has been found to shorten the G_1 phase and induce anchorage-independent growth in certain cells. Overexpression of cyclin E likewise shortens G_1 by inducing early entry into S phase. Cyclin A expression was observed to be anchorage-dependent in a variety of different fibroblast cell lines.[53] When these cells were placed in suspension culture, only cyclin A among the different cyclins was dramatically downregulated causing the suspended cells to accumulate in late G_1 phase.[54,55] In one cell line, NRK, the investigators were able to recover cells ectopically expressing cyclin A whereas other cell lines with ectopic cyclin A expression were not recoverable. Intriguingly the NRK cells with deregulated cyclin A expression were able to grow in suspension in soft agar.[53] These observations suggest that cyclin A is able to induce anchorage-independent growth. Since c-Myc expression has been linked to cyclin A and cyclin A to anchorage dependent cell growth, it was hypothesized that c-Myc induce anchorage-independent growth of Rat1a cells through deregulated cyclin A expression.[56] Indeed, Rat 1a cells in suspension displayed a dramatic reduction in cyclin A mRNA and protein levels, whereas cyclins D_1 and E remained elevated. Rat1a cells expressing ectopic c-Myc, however, showed a sustained level of cyclin A in suspension. Furthermore, several Rat 1a cell lines with inducible cyclin A expression were able to growth in soft agar. In summary, these studies suggest that cyclin A links deregulated c-Myc expression to anchorage-independent growth.

The effect of c-Myc on CDKs is unclear as well. In one study, c-Myc appears to cooperate with Ras to induce the *cdc2* promoter which does not contain a consensus Myc E-box.[57] This promoter, however, contains an NRF1 site (CGCATT) which is similar but not identical to a non-canonical Myc binding site, CGCATG, identified in in vitro studies. It was not demonstrated in this study that endogenous *cdc2* gene expression is augmented by c-Myc in vivo. In an independent study induction of c-Myc expression was shown to elevate *cdc2* gene expression.[58]

The promoter of the p53 gene was noted to contain an E-box resembling a Myc binding site.[59,60] Transient transfection experiments with p53 promoter-reporter constructs and c-Myc expression vectors resulted in transactivation. It is notable that this promoter is also transactivated by USF in transient transfection experiments. In fact, nuclear extracts demonstrated the binding of USF to this E-box, and a Myc-related shifted band was not demonstrated. Although alterations of endogenous p53 levels as a consequence of deregulated c-Myc was not studied, it is intriguing that lymphoma cell lines with high c-Myc expression tend to also have elevated p53 levels.[60] This latter observation is a correlation, which inherently does not reveal any mechanistic link between Myc and p53, and may simply reflect a functional interaction between these proteins or the pathways they elicit.

Studies of the eukaryotic translational elongation factor 2α (eIF-2α) suggest that its promoter is stimulated by c-Myc through CGCATG, an NRF1 site, which is identical to a non-canonical Myc binding site.[61,62] eIF-4E is also elevated by c-Myc. eIF-4E in turn appears to increase the translation of cyclin D_1 mRNA resulting in elevated cyclin D_1 protein level.[46] It is intriguing to note that eIF-2α appears to possess neoplastic transforming activity. NRF1 (nuclear respiratory factor) was identified as a factor that binds CGCATG which is found in a number of genes encoding the respiratory chain proteins.[63-65] Is it possible that c-Myc and NRF1 share common binding sites, but c-Myc only augments the expression of NRF1 target genes under proliferative conditions. NRF1 may serve as a constitutive positive signal in the expression of these genes. Likewise, USF and c-Myc may share common binding sites.

Overexpression of c-Myc effectively inhibits the commitment and differentiation of murine erythroleukemia (MEL) cells.[66,67] The Obinata group has examined the effects of c-Myc overexpression on selected latent period and erythroid specific genes in MEL cells.[68] The expression of latent period genes (*hsp* and *hsc70*, MER5, *Id* and *Spi-1* genes) and the erythroid-specific genes (beta-globin, glycophorin, delta-aminolevulinic acid synthase (ALAS-E), GATA-1 and erythropoietin receptor (EpoR)) in MEL cells were studied following induction of c-Myc expression. Overexpression of the c-*myc* gene positively regulated the latent period genes *hsc, hsp 70* genes and *Id* gene, whereas expression of the MER5 gene was repressed. MER5 is thought to encode an antioxidant protein. c-Myc overexpression did not affect the expression of proliferating nuclear antigen or DNA polymerase. Of the five erythroid-specific genes, c-Myc repressed the expression of glycophorin, ALAS-E and beta-globin genes, but had no effect on expression of GATA-1 or EpoR gene. These observations suggest that c-Myc differentially affects the expression of the latent period and erythroid-specific genes.

10.3 TRANSCRIPTIONAL SUPPRESSION BY c-Myc

The ability of c-Myc to repress transcription of certain genes has been recognized for some time. For example, MHC class I,[69-71] collagen genes,[5,6] *neu*,[72] LFA-173 and *myc*[74,75] itself have all been shown to be expressed in a manner inversely proportional to Myc levels. While there may be different mechanisms mediating c-Myc suppression of transcription, in vitro transcription studies suggest that c-Myc may be able to suppress initiator driven transcription.[76] Initiators are promoter elements with the consensus, YAYTCYYY (Y= pyrimidine), that allow initiation of transcription in TATA-less promoters.[77] In TATA promoters, the initiator increases transcriptional efficiency as well as allowing a precise transcriptional start site. In an assay that was dependent on the initiator and one of its binding proteins, TFII-I, c-Myc was shown to inhibit initiator dependent transcription in vitro.[76] Intriguingly, several of the genes repressed by c-Myc contain the initiator element, although it is not yet clear whether repression by c-Myc is mediated through the initiators of these promoters (Fig. 10.1).

The adenoviral major late promoter (AdMLP) contains two Myc-type E-boxes and an initiator element. The USF transcription factor was identified as the cellular factor that binds the proximal E-box and presumably plays a role in the adenoviral late gene transcription. The role of c-Myc in the adenoviral life cycle is not known, but the AdMLP has provided an ideal natural promoter to investigate c-Myc function. c-Myc is able to activate transcription of AdMLP in transient transfection experiments; however, at high input of plasmid DNA c-Myc inhibits AdMLP in an initiator dependent manner.[78] Furthermore, amino acids 122-144 in Myc box II are required for this repressive function. Promoters, such as the albumin or C/EBPα promoter, containing initiators are also repressed by c-Myc in transient transfection experiments. It appears that c-Myc also represses the cyclin D_1 promoter through its initiator sequence, although cyclin D_1 has not been shown to be repressed by Myc in other studies.[44] One of the most intriguing observations derived from these studies is that deletion of amino acids 122-144 resulted in the loss of c-Myc repression function and transforming activity but did not abrogate its transactivation function. This suggests that transcriptional repression could be important for the transforming activity of c-Myc. Whether transcriptional activation is necessary for transformation is less clear, since there have been no identified mutants of c-Myc that retain transcriptional repression function and lack transactivation activity.

Other potential cellular genes that are repressed by c-Myc include terminal deoxynucleotide transferase (Tdt) and immunoglobulin lambda gene.[79] The lambda 5 promoter initiates transcription at multiple sites containing initiator elements and confers expression in all cell types. DNA-protein complexes of Inr lambda 5:1 and Inr lambda 5:2 (sequences from two separate promoters) each contained c-Myc and Max proteins as surmised from electrophoretic mobility shift assays and specific antibodies in supershift experiments. The c-Myc and Max proteins were also found to bind the TdT initiator (InrTdT). Transient transfection assays substantiate the in vitro observations, suggesting that c-Myc may suppress the expression of these genes in vivo.

Although putative c-Myc target genes are being identified, their emergence generates more questions than the answers they attempt to provide. Which target genes are direct and which are indirect? Which genes are biologically relevant to c-Myc function in neoplasia or cell proliferation? Is transactivation by c-Myc of reporter constructs containing E-box sequences from promoters of putative target genes enough evidence to label them targets of c-Myc?

10.4 c-Myc AND APOPTOSIS

c-Myc-induced apoptosis was first recognized in studies of 32D.3 myeloid progenitor cells that are dependent on interleukin-3 (IL-3) for c-*myc* expression and growth.[27] Enforced c-*myc* expression has no effect in 32D.3 under normal growth conditions; however, in the absence of IL-3, enforced c-*myc* expression accelerates the rate of cell death and continues to drive cells into S phase. c-Myc-induced apoptosis has also been observed in serum-deprived rat 1 fibroblasts overexpressing c-Myc or expressing activated MycER.[80] The apoptotic effect of c-Myc is blocked by co-expression of Bcl-2 or exposure of serum-deprived cells to IGF-1, which may be the key survival factor in serum.[81-84] An intriguing observation recently reported is that deregulated Myc expression appears to cause Rat 1a cells to develop susceptibility to apoptosis induced by tumor necrosis factor α (TNFα) whose receptor is related to Fas.[85] This effect is blocked by the coexpression of Bcl-2. Additionally, TNFα induced apoptosis of HeLa cells is blocked by treatment of cells with antisense *myc* oligonucleotides.[86] c-Myc appears to participate in apoptosis of physiologically relevant processes as well. In a model of negative selection of thymocytes, apoptosis of A1.1 T hybridoma cells was induced by CD3/TCR activation.[87] This process was blocked by c-*myc* specific antisense oligonucleotides and required Myc-Max participation.[88]

Since the regions of c-Myc required for transcriptional regulation are also those required for apoptosis, it is surmised that c-Myc induces the transcription of genes which participate in apoptosis. Cyclin A, whose levels correlate with c-Myc expression, appears to induce apoptosis of serum-deprived cells.[47] $p34^{cdc2}$ kinase, which is a putative c-Myc target gene, has also been implicated in cytotoxic T cell (CTL) induced apoptosis.[87,89] The CTL model involves granzyme mediated apoptosis, which appears to be blocked by transient expression of human *wee*1. The *wee* 1 protein inactivates $p34^{cdc2}$ implying that granzyme induced apoptosis is mediated by $p34^{cdc2}$. p53, which is also a putative target of c-Myc, has been shown to be required for Myc-mediated apoptosis since p53 knock out fibroblasts failed to die when Myc is overexpressed in serum-deprived conditions.[90,91] These results suggest that cell cycle regulators, such as p53, cyclins and CDKs, may be critical components of Myc-induced apoptosis. However, it remains to be determined whether p53 mediates c-Myc induced apoptosis or if the presence of p53 provides a sensing mechanism that induces apoptosis when cells are inappropriately forced to enter S phase.

Ornithine decarboxylase, which is probably the best characterized of the Myc targets, also induces apoptosis when its expression is enforced.[29,92] The inhibitor of ODC, DFMO, is able to reduce Myc-mediated apoptosis. Enforced expression of ODC, however was not as effective as c-Myc in inducing apoptosis even when cells express ODC at levels equivalent to those in Myc expressing cells. Thus, ODC appears to be sufficient but not necessary for Myc-induced apoptosis. Recent results suggest that reactive oxygen species (ROS) such as hydrogen peroxide and hydroxyl radicals play a pivotal role in many forms of apoptosis. ROS induced apoptosis may provide an explanation for ODC mediated apoptosis. Polyamines produced through the activity of ODC are catabolized by oxidative pathways which may generate ROS in excess of anti-oxidants and result in apoptosis.

It is evident from many studies that deregulated c-Myc expression causes apoptosis of cells that are serum-deprived. It is not yet clear, however, whether normal regulated expression of c-Myc also induces the apoptotic pathway. Teleologically, it is reasonable to speculate that acquisition of mutations leading to deregulated Myc expression may be detrimental to the propagation of a species. Thus, unchecked enforced expression of c-Myc could be eliminated from an organism through apoptosis. Serum deprivation is an experimental artifact that may have a physiological equivalent. Cells with deregulated Myc expression would presumably grow as small tumors. When these tumors exceed the ability of blood vessels to deliver nutrients, cells at the center of these tumors would be exposed to conditions resembling serum withdrawal and therefore undergo apoptosis.[47] This is clearly an oversimplified view, but it provides rationale for further investigation of c-Myc induced apoptosis.

10.5 FUTURE DIRECTIONS AND THERAPEUTIC IMPLICATIONS

It is fair to claim that c-*myc* is in the company of two other genetic targets that are frequently altered in human cancers. The prototypical genetic target for mutations in human cancers is p53. The *ras* proto-oncogene is also frequently altered in many different human neoplastic diseases. At the laboratory bench, we look forward to better characterization of c-Myc function with specific reference to its ability to transactivate or transrepress gene expression. In looking forward we ask the following questions. What are the molecules that interact with c-Myc and regulate its function? What are the molecular underpinnings that allows c-Myc to regulate the transcriptional machinery? Are there specific co-activators for c-Myc transactivation? Are there other functions of c-Myc unrelated to its ability to regulate transcription?

At the bedside, do we expect to benefit from studies of this curious gene, its product as well as other proteins that interact with it? We believe that the benefits are already at hand. The better understanding of c-Myc in a network of complicated protein-protein interactions have already implicated the Max partner protein, Mxi-1, as a tumor suppressor. In fact, ongoing studies indicate that reintroduction of Mxi-1 into glioblastoma or prostate cancer cells, the two diseases in which significant loss of heterozygousity has been shown for Mxi-1, result in cessation of cell growth. If ectopic expression of Mxi-1 does not significantly affect normal cell function, can Mxi-1 provide an avenue for therapy of the devastating group of brain cancers termed glioblastomas? The observation that c-Myc induces apoptosis under serum-deprived conditions that may be complemented by IGF-1 suggests the possibility of exploiting these pathways to kill cancer cells with deregulated c-Myc expression. In fact, the anticancer activity of TNFα may be related to its preference for killing cells that overexpress c-Myc. Unfortunately, TNFα or cachetin has major clinical side effects making it unsuitable for therapeutic purposes. Is it possible to engineer mutant TNFα molecules that retain the ability to kill Myc expressing cells without eliciting the septic shock pathway? Are there other molecules that mimic TNFα in killing cells overexpressing c-Myc without inducing hypotensive shock?

Whatever the answers to the above questions may be, it is clear that c-Myc has been fruitful for molecular oncologists and molecular biologists at large. It will remain intriguing as a target for investigation, since c-Myc lies at the intersection between cell death and cell proliferation.

REFERENCES

1. Prendergast GC, Cole MD. Posttranscriptional regulation of cellular gene expression by the c-myc oncogene. Mol Cell Biol 1989; 9:124-34.
2. Prendergast GC, Diamond LE, Dahl D, Cole MD. The c-myc-regulated gene mrl encodes plasminogen activator inhibitor 1. Mol Cell Biol 1990; 10:1265-9.

3. Gibson AW, Ye R, Johnston RN, Browder LW. A possible role for c-Myc oncoproteins in post-transcriptional regulation of ribosomal RNA. Oncogene 1992; 7:2363-7.
4. Riccio A, Pedone PV, Lund LR, Olesen T, Olsen HS, Andreasen PA. Transforming growth factor beta 1-responsive element: closely associated binding sites for USF and CCAAT-binding transcription factor-nuclear factor I in the type 1 plasminogen activator inhibitor gene. Mol Cell Biol 1992; 12:1846-55.
5. Yang BS, Geddes TJ, Pogulis RJ, de Crombrugghe B, Freytag SO. Transcriptional suppression of cellular gene expression by c-Myc. Mol Cell Biol 1991; 11:2291-5.
6. Yang BS, Gilbert JD, Freytag SO. Overexpression of Myc suppresses CCAAT transcription factor/nuclear factor 1-dependent promoters in vivo. Mol Cell Biol 1993; 13:3093-102.
7. Schmidt A, Setoyama C, de Crombrugghe B. Regulation of a collagen gene promoter by the product of viral mos oncogene. Nature 1985; 314:286-9.
8. Slack JL, Parker MI, Robinson VR, Bornstein P. Regulation of collagen I gene expression by ras. Mol Cell Biol 1992; 12:4714-23.
9. Setoyama C, Liau G, de Crombrugghe B. Pleiotropic mutants of NIH 3T3 cells with altered regulation in the expression of both type I collagen and fibronectin. Cell 1985; 41:201-9.
10. Eizenberg O, Oren M. Reduced levels of alpha 1 (I) collagen mRNA in cells immortalized by mutant p53 or transformed by ras. Biochim Biophys Acta 1991; 1129:34-42.
11. Frankfort BJ, Gelman IH. Identification of novel cellular genes transcriptionally suppressed by v-src. Biochem Biophys Res Commun 1995; 206:916-26.
12. Hampton LL, Worland PJ, Yu B, Thorgeirsson SS, Huggett AC. Expression of growth-related genes during tumor progression in v-raf-transformed rat liver epithelial cells. Cancer Res 1990; 50:7460-7.
13. Schuur ER, Kruse U, Iacovoni JS, Vogt PK. Nuclear factor 1 interferes with transformation by nuclear oncogenes. Cell Growth Differ 1995; 6:219-27.
14. Lim A, Greenspan DS, Smith BD. Expression of alpha 2 type I collagen in W8 cells increases cell adhesion and decreases colony formation in soft agar. Matrix Biol 1994; 14:21-30.
15. Wiseman RW, Montgomery JC, Hosoi J, et al. Identification of genes associated with tumor suppression in Syrian hamster embryo cells. [Review]. Environ Health Perspect 1991; 93:105-9.
16. Eilers M, Picard D, Yamamoto KR, Bishop JM. Chimaeras of myc oncoprotein and steroid receptors cause hormone-dependent transformation of cells. Nature 1989; 340:66-8.
17. Eilers M, Schirm S, Bishop JM. The MYC protein activates transcription of the alpha-prothymosin gene. EMBO J 1991; 10:133-41.

18. Eschenfeldt WH, Berger SL. The human prothymosin alpha gene is polymorphic and induced upon growth stimulation: evidence using a cloned cDNA. Proc Natl Acad Sci USA 1986; 83:9403-7.
19. Sburlati AR, Manrow RE, Berger SL. Prothymosin alpha antisense oligomers inhibit myeloma cell division. Proc Natl Acad Sci USA 1991; 88:253-7.
20. Manrow RE, Sburlati AR, Hanover JA, Berger SL. Nuclear targeting of prothymosin alpha. J Biol Chem 1991; 266:3916-24.
21. Gaubatz S, Meichle A, Eilers M. An E-box element localized in the first intron mediates regulation of the prothymosin alpha gene by c-myc. Mol Cell Biol 1994; 14:3853-62.
22. Meichle A, Philipp A, Eilers M. The functions of Myc proteins. [Review]. Biochim Biophys Acta 1992; 1114:129-46.
23. Gaubatz S, Imho fA, Dosch R, et al. Transcriptional activation by Myc is under negative control by the transcription factor AP-2. EMBO J 1995; 14:1508-19.
24. Tavtigian SV, Zabludoff SD, Wold BJ. Cloning of mid-G(1) serum response genes and identification of a subset regulated by conditional myc expression. Mol Biol Cell 1994; 5:375-88.
25. Benvenisty N, Leder A, Kuo A, Leder P. An embryonically expressed gene is a target for c-Myc regulation via the c-Myc-binding sequence. Genes Dev 1992; 6:2513-23.
26. Benvenisty N, Ornitz DM, Bennett GL, et al. Brain tumours and lymphomas in transgenic mice that carry HTLV-I LTR/c-myc and Ig/tax genes. Oncogene 1992; 7:2399-405.
27. Askew DS, Ashmun RA, Simmons BC, Cleveland JL. Constitutive c-myc expression in an IL-3-dependent myeloid cell line suppresses cell cycle arrest and accelerates apoptosis. Oncogene 1991; 6:1915-22.
28. Auvinen M, Paasinen A, Andersson LC, Holtta E. Ornithine decarboxylase activity is critical for cell transformation. Nature 1992; 360:355-8.
29. Bello-Fernandez C, Packham G, Cleveland JL. The ornithine decarboxylase gene is a transcriptional target of c-Myc. Proc Natal Acad Sci USA 1993; 90:7804-8.
30. Pena A, Reddy CD, Wu S, et al. Regulation of human ornithine decarboxylase expression by the c-Myc.Max protein complex. J Biol Chem 1993; 268:27277-85.
31. Wagner AJ, Meyers C, Laimins LA, Hay N. c-Myc induces the expression and activity of ornithine decarboxylase. Cell Growth Differ 1993; 4:879-83.
32. Pena A, Wu S, Hickok NJ, Soprano DR, Soprano KJ. Regulation of human ornithine decarboxylase expression following prolonged quiescence: role for the c-Myc/Max protein complex. J Cell Physiol 1995; 162:234-45.
33. Packham G, Bellofernandez C, Cleveland JL. Position and orientation independent transactivation by c-Myc. Cell Molec Biol Res 1994; 40:699-706, 1994.

34. Gregor PD, Sawadogo M, Roeder RG. The adenovirus major late transcription factor USF is a member of the helix-loop-helix group of regulatory proteins and binds to DNA as a dimer. Genes Dev 1990; 4:1730-40.
35. Beckmann H, Kadesch T. The leucine zipper of TFE3 dictates helix-loop-helix dimerization specificity. Genes Dev 1991; 5:1057-66.
36. Fisher DE, Carr CS, Parent LA, Sharp PA. TFEB has DNA-binding and oligomerization properties of a unique helix-loop-helix/leucine-zipper family. Genes Dev 1991; 5:2342-52.
37. Mai S, Jalava A. C-myc binds to 5' flanking sequence motifs of the dihydrofolate reductase gene in cellular extracts - role in proliferation. Nucleic Acids Res 1994; 22:2264-73.
38. Miltenberger RJ, Sukow KA, Farnham PJ. An E-box-mediated increase in *cad* transcription at the G1/S-phase boundary is suppressed by inhibitory c-Myc mutants. Mol Cell Biol 1995; 15:2527-35.
39. Karn J, Watson J, Lowe A, Green S, Vedeckis W. Regulation of cell cycle duration by c-*myc* levels. Oncogene 1989; 4:773-87.
40. Prouty SM, Hanson KD, Boyle AL, et al. A cell culture model system for genetic analyses of the cell cycle by targeted homologous recombination. Oncogene 1993; 8:899-907.
41. Shichiri M, Hanson KD, Sedivy JM. Effects of c-myc expression on proliferation, quiescence, and the G0 to G1 transition in nontransformed cells. Cell Growth Differ 1993; 4:93-104.
42. Hanson KD, Shichiri M, Follansbee MR, Sedivy JM. Effects of c-myc expression on cell cycle progression. Mol Cell Biol 1994; 14:5748-55.
43. Jansen-Durr P, Meichle A, Steiner P, et al. Differential modulation of cyclin gene expression by MYC. Proc Natl Acad Sci USA 1993; 90:3685-9.
44. Philipp A, Schneider A, Vasrik I, et al. Repression of cyclin D1 - a novel function of myc. Mol Cell Biol 1994; 14:4032-43.
45. Daksis JI, Lu RY, Facchini LM, Marhin WW, Penn LJ. Myc induces cyclin D1 expression in the absence of de novo protein synthesis and links mitogen-stimulated signal transduction to the cell cycle. Oncogene 1994; 9:3635-45.
46. Rosenwald IB, Lazariskaratzas A, Sonenberg N, Schmidt EV. Elevated levels of cyclin D1 protein in response to increased expression of eukaryotic initiation factor 4E. Mol Cell Biol 1993; 13:7358-63.
47. Hoang AT, Cohen KJ, Barrett JF, Bergstrom DA, Dang CV. Participation of cyclin A in Myc-induced apoptosis. Proc Natl Acad Sci USA 1994; 91:6875-9.
48. Henglein B, Chenivesse X, Wang J, Eick D, Brechot C. Structure and cell cycle-regulated transcription of the human cyclin A gene. Proc Natl Acad Sci USA 1994; 91:5490-4.

49. Yamamoto M, Yoshida M, Ono K, et al. Effect of tumor suppressors on cell cycle-regulatory genes: RB suppresses p34cdc2 expression and normal p53 suppresses cyclin A expression. Exp Cell Res 1994; 210:94-101.
50. Yoshizumi M, Hsieh CM, Zhou F, et al. The ATF site mediates downregulation of the cyclin A gene during contact inhibition in vascular endothelial cells. Mol Cell Biol 1995; 6:3266-72.
51. Desdouets C, Matesic G, Molina CA, et al. Cell cycle regulation of cyclin A gene expression by the cyclic AMP-responsive transcription factors CREB and CREM. Mol Cell Biol 1995; 6:3301-9.
52. Nakamura T, Okuyama S, Okamoto S, Nakajima T, Sekiya S, Oda K. Down-regulation of the cyclin A promoter in differentiating human embryonal carcinoma cells is mediated by depletion of ATF-1 and ATF-2 in the complex at the ATF/CRE site. Exp Cell Res 1995; 216:422-30.
53. Guadagno TM, Ohtsubo M, Roberts JM, Assoian RK. A link between cyclin A expression and adhesion-dependent cell cycle progression [published erratum appears in Science 1994 Jan 28; 263(5146):455]. Science 1993; 262:1572-5.
54. Guadagno TM, Assoian RK. G1/S control of anchorage-independent growth in the fibroblast cell cycle. J Cell Biol 1991; 115:1419-25.
55. Han EK, Guadagno TM, Dalton SL, Assoian RK. A cell cycle and mutational analysis of anchorage-independent growth: cell adhesion and TGF-beta 1 control G1/S transit specifically. J Cell Biol 1993; 122:461-71.
56. Barrett JF, Lewis BC, Hoang AT, Alvarez JRJ, Dang CV. Cyclin A links c-Myc to adhesion-independent cell proliferation. J Biol Chem 1995; 270:15923-5.
57. Born TL, Frost JA, Schonthal A, Prendergast GC, Feramisco JR. C-myc cooperates with activated ras to induce the cdc2 promoter. Mol Cell Biol 1994; 14:5710-8.
58. Kim YH, Buchholz MA, Chrest FJ, Nordin AA. Up-regulation of c-myc induces the gene expression of the murine homologues of p34(cdc2) and cyclin-dependent kinase-2 in T lymphocytes. J Immun 1994; 152:4328-35.
59. Reisman D, Elkind N, Roy B, Beamon J, Rotter V. c-Myc transactivate the p53 promoter through a required downstream CACGTG motif. Cell Growth Differ 1993; 4:57-65.
60. Roy B, Beamon J, Balint E, Reisman D. Transactivation of the human p53 tumor suppressor gene by c-Myc/Max contributes to elevated mutant p53 expression in some tumors. Mol Cell Biol 1994; 14:7805-15.
61. Rosenwald IB, Rhoads DB, Callanan LD, Isselbacher KJ, Schmidt EV. Increased expression of eukaryotic translation initiation factors eIF-4E and eIF-2 alpha in response to growth induction by c-myc. Proc Natl Acad Sci USA 1993; 90:6175-8.

62. Blackwell TK, Huang J, Ma A, et al. Binding of myc proteins to canonical and noncanonical DNA sequences. Mol Cell Biol 1993; 13:5216-24.
63. Chau CM, Evans MJ, Scarpulla RC. Nuclear respiratory factor 1 activation sites in genes encoding the gamma-subunit of ATP synthase, eukaryotic initiation factor 2 alpha, and tyrosine aminotransferase. Specific interaction of purified NRF-1 with multiple target genes. J Biol Chem 1992; 267:6999-7006.
64. Virbasius CA, Virbasius JV, Scarpulla RC. NRF-1, an activator involved in nuclear-mitochondrial interactions, utilizes a new DNA-binding domain conserved in a family of developmental regulators. Genes Dev 1993; 7:2431-45.
65. Virbasius JV, Scarpulla RC. Activation of the human mitochondrial transcription factor A gene by nuclear respiratory factors: a potential regulatory link between nuclear and mitochondrial gene expression in organelle biogenesis. Proc Natl Acad Sci USA 1994; 91:1309-13.
66. Takada S, Obinata M. Overexpression of c-Myc inhibits the appearance of a specific DNase I hypersensitive site in the beta-globin chromatin in murine erythroleukemia cells. Jpn J Cancer Res 1991; 82:376-9.
67. Takada S, Yamamoto T, Ohmori Y, Matsui Y, Obinata M. c-Myc interferes with the commitment to differentiation of murine erythroleukemia cells at a reversible point. Jpn J Cancer Res 1992; 83:61-5.
68. Shoji W, Ohmori Y, Obinata M. c-Myc selectively regulates the latent period and erythroid-specific genes in murine erythroleukemia cell differentiation. Jpn J Cancer Res 1993; 84:885-92.
69. Bernards R, Dessain S, Weinberg R. N-*myc* amplification causes down-modulation of MHC class I antigen expression in neuroblastoma. Cell 1986; 47:667-74.
70. Lenardo M, Rustgi A, Schievella A, Bernards R. Suppression of MHC Class 1 gene expression by N-myc through enhancer inactivation. EMBO J 1989; 8:3351-5.
71. van't Veer LJ, Beijersbergen RL, Bernards R. N-myc suppresses major histocompatibility complex class I gene expression through down-regulation of the p50 subunit of NF-kappa B. EMBO J 1993; 12:195-200.
72. Suen T-C, Hung M-C. c-myc reverses neu-induced transformation morphology by transcriptional repression. Mol Cell Biol 1991; 11:354-62.
73. Inghirami G, Grignani F, Sternas L, Lombardi L, Knowles DM, Dalla-Favera R. Down-regulation of LFA-1 adhesion receptors by C-myc oncogene in human B lymphoblastoid cells. Science 1990; 250:682-6.
74. Penn LJ, Brooks MW, Laufer EM, Land H. Negative autoregulation of c-myc transcription. EMBO J 1990; 9:1113-21.

75. Penn LJ, Brooks MW, Laufer EM et al. Domains of human c-myc protein required for autosuppression and cooperation with ras oncogenes are overlapping. Mol Cell Biol 1990; 10:4961-6.
76. Roy AL, Carruthers C, Gutjahr T, Roeder RG. Direct role for Myc in transcription initiation mediated by interactions with TFII-I. Nature 1993; 365:359-61.
77. Smale ST, Baltimore D. The "initiator" as a transcription control element. Cell 1989; 57:103-13.
78. Li LH, Nerlov C, Prendergast G, MacGregor D, Ziff EB. c-Myc represses transcription in vivo by a novel mechanism dependent on the initiator element and Myc box II. EMBO J 1994; 13:4070-9.
79. Mai S, Martensson IL. The c-myc protein represses the lambda 5 and TdT initiators. Nucleic Acids Res 1995; 23:1-9.
80. Evan GI, Wyllie AH, Gilbert CS, et al. Induction of apoptosis in fibroblasts by c-myc protein. Cell 1992; 69:119-28.
81. Fanidi A, Harrington EA, Evan GI. Cooperative interaction between c-myc and bcl-2 proto-oncogenes. Nature 1992; 359:554-6.
82. Wagner AJ, Small MB, Hay N. Myc-mediated apoptosis is blocked by ectopic expression of Bcl-2. Mol Cell Biol 1993; 13:2432-40.
83. Bissonette R, Echeverri F, Mahboubi A, Green D. Apoptotic cell death induced by c-*myc* is inhibited by *bcl*-2. Nature 1992; 359:552-4.
84. Harrington EA, Bennett MR, Fanidi A, Evan GI. C-myc-induced apoptosis in fibroblasts is inhibited by specific cytokines. EMBO J 1994; 13:3286-95.
85. Klefstrom J, Vastrik I, Saksela E, Valle J, Eilers M, Alitalo K. c-Myc induces cellular susceptibility to the cytotoxic action of TNF-alpha. EMBO J 1994; 13:5442-50.
86. Janicke RU, Lee FH, Porter AG. Nuclear c-Myc plays an important role in the cytotoxicity of tumor necrosis factor alpha in tumor cells. Mol Cell Biol 1994; 14:5661-70.
87. Shi Y, Glynn JM, Guilbert LJ, Cotter TG, Bissonnette RP, Green DR. Role for c-myc in activation-induced apoptotic cell death in T cell hybridomas. Science 1992; 257:212-4.
88. Bissonnette RP, McGahon A, Mahboubi A, Green DR. Functional Myc-Max heterodimer is required for activation-induced apoptosis in T cell hybridomas. J Exp Med 1994; 180:2413-8.
89. Shi L, Nishioka WK, Th'ng J, Bradbury EM, Litchfield DW, Greenberg AH. Premature p34cdc2 activation required for apoptosis. Science 1994; 263:1143-5.
90. Wagner AJ, Kokontis JM, Hay N. Myc-mediated apoptosis requires wild-type p53 in a manner independent of cell cycle arrest and the ability of p53 to induce p21waf1/cip1. Genes Dev 1994; 8:2817-30.
91. Hermeking H, Eick D. Mediation of c-myc-induced apoptosis by p53. Science 1994; 265:2091-3.
92. Packham G, Cleveland JL. Ornithine decarboxylase is a mediator of c-myc-induced apoptosis. Mol Cell Biol 1994; 14:5741-7.

INDEX

Page numbers in italics denote figures (f) and tables (t).

V

W

Y

Z

Medical Intelligence Unit

Available and Upcoming Titles

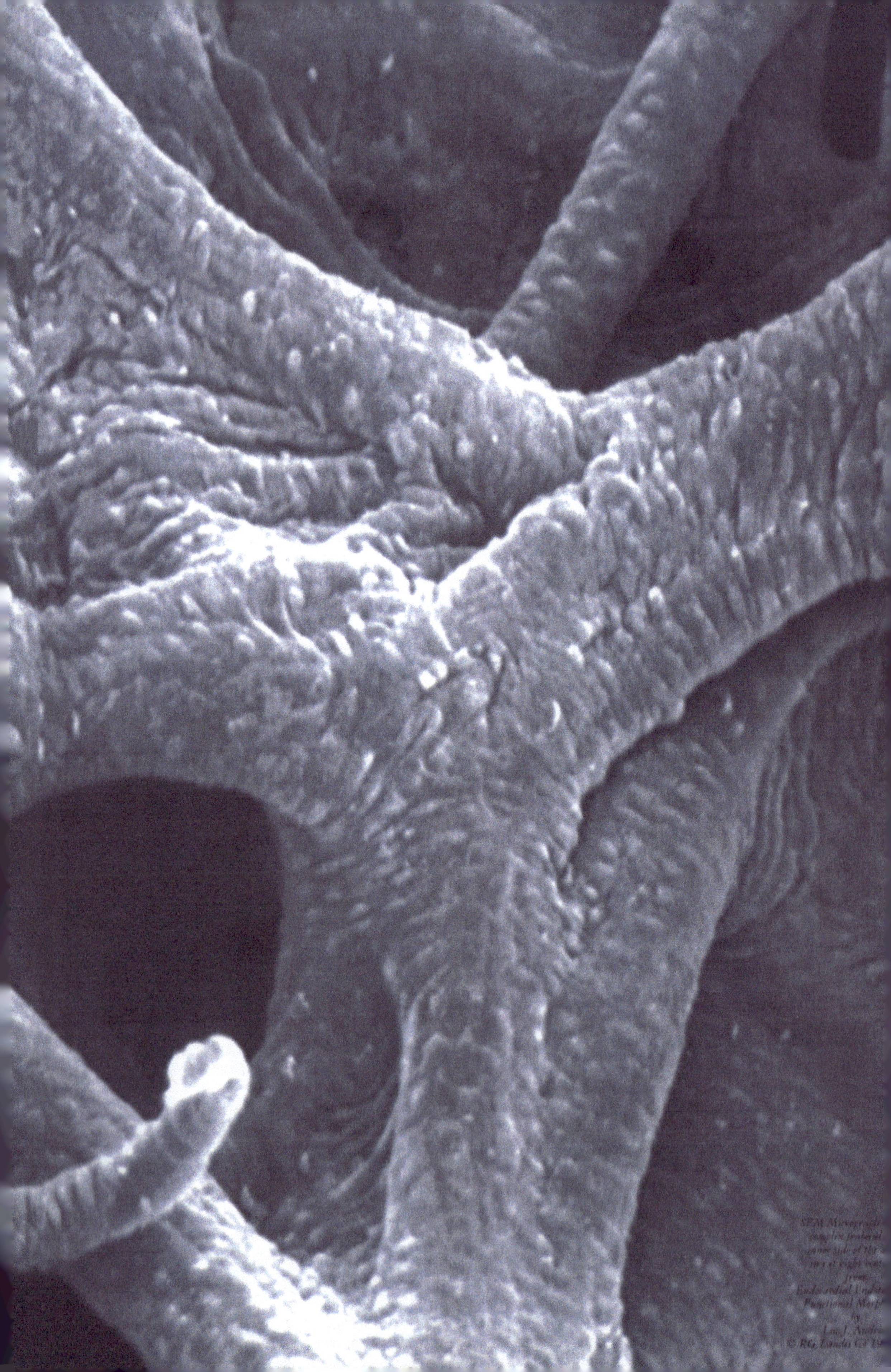
SEM Micrograph
from
Endocardial Endo
Functional Morph
by
Leo J. Andr
© RG Landes Co 19

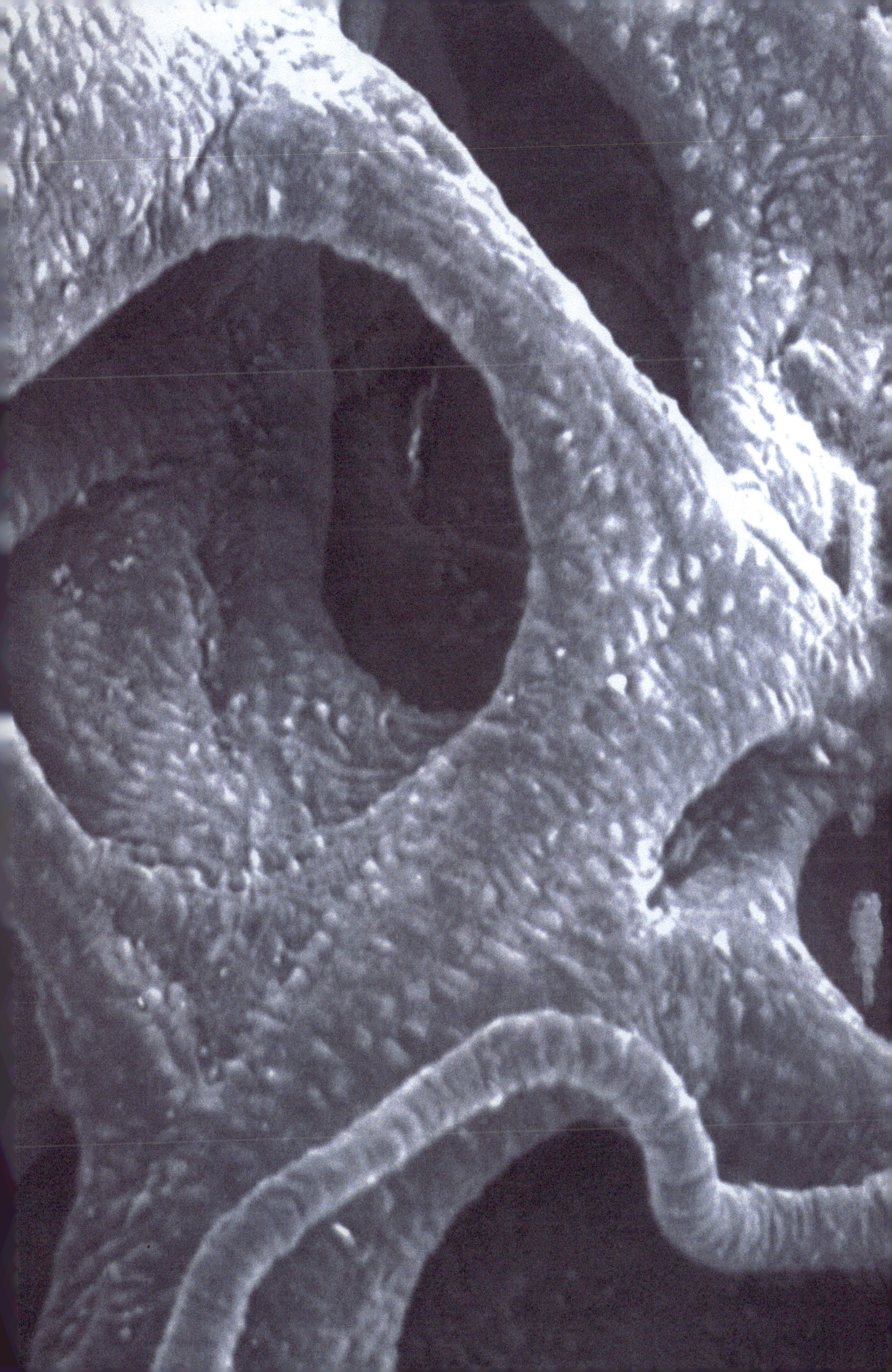

MIX
Papier aus verantwortungsvollen Quellen
Paper from responsible sources
FSC® C105338

If you have any concerns about our products,
you can contact us on
ProductSafety@springernature.com

In case Publisher is established outside the EU,
the EU authorized representative is:
Springer Nature Customer Service Center GmbH
Europaplatz 3, 69115 Heidelberg, Germany

Printed by Libri Plureos GmbH
in Hamburg, Germany